# Hacking Wireless Access Points

# Hacking Wireless Access Points

## Cracking, Tracking, and Signal Jacking

**Jennifer Ann Kurtz**
*Information Assurance Affiliate Faculty at Regis University*

**Richard Kaczmarek, Technical Editor**

AMSTERDAM • BOSTON • HEIDELBERG • LONDON
NEW YORK • OXFORD • PARIS • SAN DIEGO
SAN FRANCISCO • SINGAPORE • SYDNEY • TOKYO
Syngress is an imprint of Elsevier

SYNGRESS.

Syngress is an imprint of Elsevier
50 Hampshire Street, 5th Floor, Cambridge, MA 02139, United States



**British Library Cataloguing-in-Publication Data**
A catalogue record for this book is available from the British Library

**Library of Congress Cataloging-in-Publication Data**
A catalog record for this book is available from the Library of Congress

ISBN: 978-0-12-805315-7

For Information on all Syngress publications visit our website at https://www.elsevier.com

*Publisher:* Todd Green
*Acquisition Editor:* Chris Katsaropoulos
*Editorial Project Manager:* Anna Valutkevich
*Production Project Manager:* Priya Kumaraguruparan
*Cover Designer:* Mark Rogers

Typeset by MPS Limited, Chennai, India

To my children, grandchildren, and all who would use technology responsibly.

# Contents

# About the Author

**Jennifer Kurtz** is a technology project manager, business development consultant, educator, and writer, currently focused on information security, privacy, organizational resilience, and marketing strategy. She has held appointments at Purdue University and Ball State University and currently teaches and develops graduate courses in information assurance at Regis University. Her work in telecommunications includes leading statewide broadband infrastructure and eGovernment initiatives as Indiana's director of eCommerce, building and managing the telecommunications infrastructure for Delco Remy International, and coauthoring a 10-year strategic plan for the US Department of the Treasury. She wrote the chapter on data leakage prevention for the American Bar Association's 2011 best-selling book, *The Data Breach and Encryption Handbook*, in addition to other publications and information assurance blogs. Her degrees are from The American University and Anderson University.

# Preface

As a child in rural Ohio, I marveled over descriptions in the weekly *Junior Scholastic* of President Eisenhower's proposed Interstate Highway System that would contact the east and west coasts for high-speed, unimpeded, toll-free travel. Innovative cloverleaf exit and entrance ramps, with their ingenious reverse direction to change direction configuration, would encourage safe throughput by eliminating right-angle turns and sudden stops: Cars, trucks, and motorcycles would merge easily with others. I was eager to cruise this highway and enjoy its streamlined efficiency without all the speed traps and slowdowns in small town that vexed my father and delayed family and vacation visits.

And that's what we were promised with the information super highway as well: free high-speed communications to the whole world of knowledge and instantaneous communication. On the one hand, success can be measured by global adoption rates of Internet and smartphone use that are pushing 67% and 43%, respectively (for those between the ages of 18 and 35, the numbers are much higher).[1] But we're also encountering similar disappointing realities in digital space as in physical space: disconnects in end-to-end movement, constant repair, random events—and the cyber equivalent of drunken driving, road rage, recklessness, and accidents. We appreciate the need for commonly understood and practiced rules of the road, protective gear, situational awareness, and defensive navigation.

In the pursuit of speed, convenience, low cost, and entertainment, however, we have inadvertently created pathways for opportunistic predators who do

[1] According to the early 2016 survey report from the Pew Research Center, Internet use among those between the ages of 18 and 35 in the "advanced economies" surveyed ranges between 96% and 100%; in 50% of the "emerging economies" surveyed, Internet use was at least 75%. Smartphone use lags Internet use, not surprisingly, with only 7 of the 11 advanced economies surveyed showing use greater than 90% in this age group. Still, the degree of penetration is impressive. Jacob Poushter (February 22, 2016), "Smartphone Ownership and Internet Usage Continues to Climb in Emerging Economies," *Pew Research Center.* Retrieved from http://www.pewglobal.org/2016/02/22/smartphone-ownership-and-internet-usage-continues-to-climb-in-emerging-economies/.

not believe in behavioral limits or respect for the rights (and property) of others. In this book I've attempted to show from a mobile, wireless perspective where these bad actors can find vulnerabilities and seek to take advantage. By looking at real-world attack scenarios, we can begin to think like them: What would attackers do? They have many tools and choices—as do we. We can operate our mobile devices more safely, we can navigate our wireless highways more thoughtfully, and we can purchase products more judiciously. We can use what we know and learn what we don't already.

The first two chapters describe the general historical and technical background for wireless access points (WAPs): definitions, specifications, standards, and emerging trends. The third chapter looks at hacker motivations and different categories of attacks. The next chapters focus on different operational environments for WAPs: individual consumer (Chapter 4); commercial/industrial (Chapter 5); medical and health care (Chapter 6); civilian government (Chapter 7); non-civilian government for public safety, emergency management, and national security (Chapter 8). Chapters 4 through 7 include at least one actual attack scenario and a list of takeaways. Due to the greater caution taken with attack scenarios that occur within the non-civilian government environment, no attack scenario is outlined here. Chapter 9 summarizes observations and takeaways, and delivers a call to action and makes an appeal for the responsible use of technology. The Appendix contains a glossary of WAP-related terms and attack tree diagrams.

This exploration covers a broad territory: The drill down into each focus environment could become a book in its own right. It is intended to be suggestive, disturbing, and accessible, as any good call to action ought to be. I hope it brings readers to the conclusion that it is time for them to get cracking, recognize hacking, and impede signal jacking.

# Acknowledgments

Special thanks to my technical editor and long-time mentor, Richard Kaczmarek, who described his role in reviewing my drafts as "helping you find your voice and say what you believe to be true and important." Any misstatements are my own. He and Elsevier editorial project manager, Anna Valutkevich, encouraged me to come through the rabbit holes of research and commit words to paper—and then let them go and continue. I would also like to express my appreciation to the scholars who have given me opportunities to develop knowledge about technology and share it with others: Dan Likarish and Bob Bowles (Regis University), Professor Eugene "Spaf" Spafford (Purdue University), Professor Stephan Jones (Ball State University), Roland Cole (Sagamore Institute for Policy Research), and Amy Conrad Warner (Indiana University-Purdue University at Indianapolis). Thank you to Chris Katsaropoulos (Elsevier Publishing) for taking a chance with a first-time book author.

And to all who take the time to peruse these chapters, consider how technology can be used and abused, and choose to thwart the latter: Courage! We all need to think like a black hat and behave like a white hat.

CHAPTER 1

# Wireless Technology Overview

## THE WONDER OF WANDERING SIGNALS

Alien Harry Solomon (played by French Stewart) experienced minor convulsions whenever the chip implanted in his brain started to transmit an "Incoming message from the Great Big Head" in the TV show *3rd Rock from the Sun*. How 20th century! The March 2012 South by Southwest (SxSW) Interactive Conference in Austin, TX, introduced homeless human beings as wireless access points. Encountering challenges with your wireless signal? Just "log onto" a human wearing a T-shirt that says, "I'm a 4 G hotspot." No alien force needed. How 21st century!

Outrageous? Yes. These human wireless access points anticipated receiving $2 per 15-minute signal negotiation (the recommended donation). How many of us would choose to become wireless access points for $8 an hour? How many of us would mindlessly opt to become transmitters for convenience? How many of us already have?

Any device that transmits and receives data wirelessly is, by definition, a wireless access point or transceiver. Consider the number of transmitters we use. There are, of course, the usual suspects: mobile phones, laptops, US passport cards (since 2008), digital navigation systems, printers. On top of that, applications for Bluetooth and Radio Frequency Identification (RFID) technologies continue to multiply: athletic shoes, heart rate monitors, fitness sensors, cameras, printers, headsets, and so on. Gartner, Inc. forecast a 30% increase in the number of connected things from 2014 and 2015, to 4.9 billion, with another 20 billion coming online by 2020.[1] Even more exuberantly, Juniper Research published a report forecasting that 2020 would actually see more than 38 billion attached units.[2] Cisco has suggested 50 billion attachments, generating some $8 trillion worldwide in value at stake through "innovation and revenue ($2.1 trillion), asset utilization ($2.1 trillion), supply chain and logistics ($1.9 trillion), employee productivity improvements ($1.2 trillion), and enhanced customer and citizen service ($700 billion)."[3] The Internet of



Hacking Wireless Access Points. DOI: http://dx.doi.org/10.1016/B978-0-12-805315-7.00001-2

Things (IoT), possibly even the Internet of Everything (IoE), belongs among the top 10 critical IT trends into the next decade.

Industrial, governmental, and individual applications are proliferating—as are concerns about how to protect applications and users—especially given links of connections and ad hoc data connections created unintentionally. It is one thing to transmit information about one's own identity or behavior. It is more concerning when we transmit information about or from others, especially if that transmission is unconscious. By essentially leaving our personal or our organization's Internet access open—for example, by not implementing sufficiently strong security on wireless routers, by leaving Bluetooth devices in a discoverable mode, or by allowing bridging traffic between networks—we can enable irresponsible behavior by others who use our legitimate, but open, wireless subscription services. We not only expose ourselves and our organizations, but we become the "legitimate" wireless access point for another who may be pursuing illicit Internet activities, activities rendered more anonymous by piggybacking on our valid signals.

Mobile devices are widely characterized as another attack surface. This understates the reality: Mobile devices are frequently attach surfaces; information attaches to them informally in ad hoc network configurations; devices can then tunnel in and attach to more formal information networks. The conceptualization of an attack surface does not fully capture the complications inherent in mobile devices. The term "surface" resonates with more two-dimensional perimeter models, in which the internal or trusted environment is clearly differentiated from the external or untrusted environment. Mobile devices behave more like skin, however: porous, capable of two-way transmissions, and composed of multiple layers. They operate in free space, and so are more elusive and more pervasive than just another attack surface.

And like skin, mobile devices and wireless systems generally can be compromised at different levels and yet can be made more resilient to certain environmental conditions than some fixed, hardwired devices and systems. Mobile devices can also be observed for early symptoms indicating that the entire system of information connectivity, wired and wireless, is gone amiss. They gather information as well as communicate it. And as with skin, wireless security begins with good, basic hygiene.

Just because it is wireless does not mean it is unconnected, wherein lie risk and opportunity. The "radio" aspect of wireless is where some trouble begins. Adapted from the Latin root for "radiate," the word itself captures one-half of the essential nature of radios and other devices that incorporate radio wave technology: They broadcast or transmit electromagnetic signals that announce "I'm here." Radio is a chatty medium. In terms of security appetite, it tends to the permissive/promiscuous side, rather than the prudent/paranoid.

The other piece of the essential nature of wireless devices is that they can also receive electromagnetic signals. In that transmission/reception pairing is the opportunity for interference, whether from changes in power line signals (e.g., when turning on the light in the TV room interrupts the signal from the exterior-mounted antenna), competing frequencies (e.g., the crosstalk heard over cell phones), eavesdropping, interception, or loss of signal strength. Such interference can be incidental, annoying, ephemeral, or malicious. In the latter case, interference can be deliberately used to capture or divert signal content, compromise signal capacity, discover vulnerabilities in the transmitting system, and otherwise interrupt communication signal flow, including the sending of misleading feedback or communication content.

I remember as a child in the mid-1950s that our black-and-white Magnavox TV could not compete with the electric Mixmaster in the kitchen. Mashed potato preparation meant that Lassie's canine grace was reduced to audio static and visual turbulence. Portable TVs were not easily moved without a TV cart, and frequent adjustments of the rabbit ear antennae were required. My sisters and I took turns changing stations, acting as remote controler for our parents and grandparents while they checked for programs being broadcast in color. And receiving a transistor radio for Christmas meant that I could, magically, enjoy music wherever I walked (after I found static-free stations).

Now in the mid-2010s, my granddaughter casually picks up my smartphone and downloads games, shares photos, and streams music videos without questioning that things should be otherwise. Meanwhile, I worry about whether she understands that some games will volunteer information about my phone and its configuration, that not all websites are safe, and that her musical preferences might be shared with marketers (along with my business contact information). Were I using my phone to access the university's network or my clients' financial information, I would be even more concerned. Were my granddaughter older, I would be worried about smart cars (training smarter drivers seems more desirable to me) and smart highways.

Closet Luddite though I may be, I am nonetheless delighted that the technology that enables the portability of devices and the mobility of information can promote productivity, learning, medical successes, public safety, entertainment, and general social value. At the same time, it's evident that the technology to subvert those good outcomes may be developing at an even faster pace. To paraphrase Symbolist poet William Blake, the road of wireless access leads to the palace of rueful wisdom for the unprepared cyber traveler.

This book is about preparing our organizations, our governments, and ourselves for making good connectivity choices. We need to develop thicker (or at least more resilient) skin, as it were. We can define how, when, and what we want to communicate to others through shared, especially wireless, connections. In many cases, we can use technology to implement those decisions more efficiently and, as organizational leaders, to substantiate that we have made acceptable, reasonable, and responsible decisions to protect our constituents, clients, and business partners. Protection will not be perfect; those with malicious intent, like Deuteronomy's poor, "will never cease out of the land." We can, however, become less successful targets for hacking. To do this, we must determine first to understand what our security negotiating position is, collect only what information we need, communicate only what data we must, use only the wireless services that support our current activities, and protect signals and systems prudently (but not paranoically).

## WIRELESS DEVICES, SIMPLIFIED

Our wireless devices are wonders to behold. Consider a typical smartphone. It is first and foremost a computer with an operating system. As with any computer, the rules of security hygiene apply: Keep your operating system up to date, use anti-virus (AV) software, and do not use easily guessable passwords.

The smartphone typically supports one or two technologies associated with the mobile carrier, a Wi-Fi connection so the smartphone can use a home network or commercial hotspot (in order to save minutes or megabytes associated with mobile carrier limits), and a Bluetooth connection to support a wireless headset and hands-free communications while operating an automobile or treadmill.

How does a smartphone vendor know what features to provide in its products? It reviews a set of technology standards with its mobile carrier partners to determine what features are mandatory for the product to function and what features provide a competitive advantage to the vendor. We will talk about technology standards in the next section.

With few exceptions, users typically refuse to pay more for devices with additional security features. Vendors and carriers may have security features turned off or set to minimum capabilities out of the box so that naïve users do not complain about being unable to access certain sites or features. This means that we as individuals and corporate or governmental staff must determine how to use mobile technology responsibly and in compliance with prevailing policy in our respective environments. We can determine what security features our smartphone has and what settings provide the appropriate security for our devices.

## THE TELECOMMUNICATIONS STANDARDS LANDSCAPE

Telecommunications standards are important in that they provide a mechanism that allows interoperability—a common vocabulary, in effect—and thus enables competition among device manufactures and carriers. Without standards, a user may conceivably need to have a device for each network or function within a network.

There are a number of international, regional, and country-specific standards bodies and consortia that contribute to the standards making process. Since the technology is constantly evolving, a key role of the standards organizations and consortia is to specify the functionality of existing and future standard releases.

At the international level exists the International Standards Organization (ISO) and International Telecommunications Union (ITU). The ISO is a nongovernmental body composed of the standards organization of member countries. The ITU specializes in information and communications technologies (ICTs), and has governmental (voting) members as well as nonvoting members such as carriers, equipment and software vendors, and international and regional telecommunications organizations.

Regional standards organizations include European Telecommunications Standards Institute (ETSI) and the Alliance for Telecommunications Industry Solutions (ATIS) in North America. ETSI is responsible for the standardization of ICTs within Europe; meanwhile, ATIS performs a similar function for the United States. Within the United States, the National Institute of Standards and Technology (NIST) is the federal agency that works with the telecommunications industry to articulate applied technology measurements and standards.

A number of consortia are responsible for development of telecommunications technical standards. These include the Third Generation Partnership Project (3GPP), the Internet Engineering Task Force (IETF), and the Institute of Electrical and Electronics Engineers (IEEE). 3GPP is responsible for a number of mobile (cellular) technologies. IETF is responsible for a number of the protocols used in both wireless and wireline networks. IEEE is responsible for a number of local area network (LAN) technologies, including Wi-Fi and Bluetooth.

All standards organizations and consortia consist of subgroups that focus on specific aspects of the technologies, such as functionality, provisioning and operational considerations, interoperability testing, and security. Organizations like the IEEE can be assisted by forums devoted to specific technologies, such as the Wi-Fi Alliance.

Specific organizations and consortia of interest to the exploration of protecting wireless access points from hacking have established rules of engagement based on a shared understanding of key factors (i.e., processes or layers) at play in interconnecting any two communicating devices.

## TELECOMMUNICATIONS RULES OF ENGAGEMENT

The ISO has provided a key standard for use in our discussions: the Open Systems Interconnection (OSI) model. The model is composed of seven layers:

1. Physical: The mechanisms for a device to attach to the media and to communicate over the media.
2. Link: The mechanisms for a device to identify (address) other devices on the media and to communicate with these devices.
3. Network: The mechanisms for a device to identify (address) other devices on different networks and to communicate with these devices.
4. Transport: The mechanisms to provide different quality of service (QoS) capabilities to applications, such as packet loss or recovery.
5. Session: The mechanisms to identify the start and end of a conversation between two devices.
6. Presentation: The mechanisms to help applications present the context of the messages being sent (e.g., converting a web page to fit on a smartphone screen).
7. Application: The programs and functions that use the communications.

When an application on Device A wants to send a message to Device B, it sends the message to the presentation layer. The presentation layer may make format changes to the message and break the message into packets for transmission. The presentation layer then sends these packets to the session layer. The session layer determines if a conversation is already in progress with Device B; if no conversation is in progress, the session layer will establish a session with Device B. It then sends the packets of the message to the transport layer.

Based on the application, the transport layer determines the protocol used to send the packets to the destination. Device B may expect packets in order with no loss, or may be willing to accept packets in any order with packet loss. The transport layer may also determine whether the packets need to be resized to send over the network. These packets are then passed to the network layer. The network layer applies the appropriate network addresses (typically Internet Protocol or IP addresses) to the packets and sends the packets to the link layer. The link layer applies the appropriate link addresses (typically Media Access Control or MAC addresses) to the packets and transmits the packets over the physical layer.

The advantage of OSI layering approach is that one layer can be modified and upgraded to take advantages of newer technology without impacting the functioning of the other layers. This is why web browsing works over cable connections, Wi-Fi, and on mobile phones. The disadvantage of the OSI layering approach is that each layer has security risks (e.g., phishing attack at an email application, telephone number spoofing at the session layer, transmitting device identification information in the open at the link layer). We will discuss these security risks in later chapters.

The ITU is celebrating its sesquicentennial. Established in 1865 as the International Telegraph Union, it is the oldest international organization and, since 1947, has been the United Nations agency that specializes in ICTs. ITU is "committed to connecting all the world's people" and performs by allocating radio spectrum and satellite orbits, developing technical interconnectivity standards, and promoting ICT benefits among underserved populations.

Pertinent to our discussions, ITU has proposed a defense-in-depth approach for communications systems that covers three distinct planes of network activity—management, control, and end-user—that must be protected to achieve eight security objectives at the different layers of the OSI stack, which it groups as infrastructure, services, and applications. In its Recommendation X.805 for security architecture, the objectives identified are access control, authentication, nonrepudiation, data confidentiality, communication security, data integrity, availability, and privacy.[4]

Communications media (e.g., copper, cable, satellite, optical, radio) and network topologies differ inherently in how the identified security objectives can be realized. Signals transmitted over copper, for example, are more conducive to communication security without additional security mechanisms than are signals transmitted via radio waves. Network topologies also differ inherently in how well they support end-to-end security in communications along all dimensions. Point-to-point connectivity, for example, is highly conducive to achieving privacy objectives, but is vulnerable to single point of failure issues, thus compromising the availability objective unless redundancy is provided.

Anyone who has traveled to Europe with a US-based mobile phone has had the experience of choosing how (or whether) to manage service coverage while there to accommodate GSM service preferences. Interoperability of systems that span governmental jurisdictions is abetted by cooperation between standards organizations and commercial industry. ETSI is one of the key certifying bodies for wireless standards like GSM. Other kinds of global services under development—smart grid, for example—are coordinated through standards organizations that include NIST (USA) and the European Union's Smart Grid Coordination Group (SG-CG). The latter represents private sector standards organizations: ETSI, the European Committee for Standardization (CEN), and the European Committee for Electrotechnical Standardization (CENELEC).[5]

## WIRELESS COMMUNICATIONS RULES OF ENGAGEMENT

Basic wireless functionality is enabled when a radio device, at once receiver and transmitter, announces its presence to make a connection. Along the wireless spectrum, frequencies vary in wavelength, signal intensity, error rate, mobility, and broadcast cone. How, then, is connection negotiated? As mentioned earlier interference from other systems and devices can be an issue in wireless systems, as can problems with interoperability. The latter is one of the challenges encountered by ad hoc teams of public safety or rescue personnel from different jurisdictions (in the United States, that would mean from different counties or states, as in a firefighting or natural disaster situation) when they assemble under emergency conditions and experience incompatible radio signaling.

Standards organizations convene technical researchers and practitioners to establish common approaches, as do some industry consortia. The three main groups of interest with respect to wireless communications are 3GPP, IETF, and IEEE.

3GPP is the organization responsible for the standards associated with GSM, UMTS, and LTE technologies, as well as the future "5 G" technology International Mobile Telecommunications (IMT) 2020. According to its website,[6] 3GPP unites seven standards development organizations (including ATIS and ETSI) and provides them with an environment to produce 3GPP technology reports and specifications. The focus of 3GPP is on cellular telecommunications network technologies, including radio access networks, core transport networks, and the services using these technologies. 3GPP also provides specifications for connecting nonradio (i.e., wireline) access to the core network and Wi-Fi internetworking.

The IETF is an international organization open to anyone interested in understanding or progressing the architectures and technologies used in the Internet.[7] Working group topics and discussions are mostly handled by mailing lists. Face-to-face meetings occur three times a year (typically in the Americas, Europe, and Asia); participants who cannot attend the meetings in person can access working group sessions through the Internet. The IETF is responsible for the standards associated with the IP suite of network, transport, and session layers. Many of the protocols developed by the IETF assumed a level of trust that could be exploited by hackers. Determining how to secure the protocols currently in use is one of the many efforts currently being worked on by the IETF.

The IEEE is the world's largest organization for technology professionals, with more than 426,000 members (of whom US professionals represent less than half) and a presence in at least 160 countries. It was established in 1963 from the joining together of the American Institute of Electrical Engineers and the Institute of Radio Engineers, respectively founded in 1884 and 1912.[8]

The Wi-Fi Alliance, founded by a group of technology companies that helped pioneer higher speed wireless networking with the IEEE 802.11b specification (3Com; Aironet, a Cisco acquisition; Harris Semiconductor, now Intersil; Lucent, now Alcatel-Lucent; Nokia; and Symbol Technologies, now Motorola),[9] has grown to include more than 600 leading global technology companies whose shared vision is "Connecting everyone and everything, everywhere."[10] With some two billion Wi-Fi devices sold in 2013 and at least 25% of homes worldwide Wi-Fi-enabled, the Alliance continues to build on the IEEE 802.11i foundational work for securing transmissions by incorporating both encryption and authentication mechanisms into the replacement, partial security solutions to wireless equivalent privacy (WEP) for Wi-Fi protected access, WPA and WPA2.

WEP is part of the original 1997 802.11 standard for wireless networking. The increase in processing speeds, which averaged about 25% annually between 1997 when the standard was first ratified until 2002, has a twofold significance. On the one hand, more complex encryption algorithms can be implemented without an unacceptable compromise in performance. On the other hand, tools for breaking encryption are also more sophisticated, fast, and accessible. An encryption algorithm deemed acceptable in 1997 is now considered too easily broken.[11]

Additional security solutions should be layered into the nonresidential Wi-Fi environment, in particular, to address casual (or malicious) signal detection from the access points and the connecting devices, as well as the unauthorized deployment of wireless access points (also known as rogue access points). Given the existence of easily accessible tools for identifying information broadcast about the service set identifier (SSID) and MAC address, organizations should determine whether applying security controls to those potential exposure points will deliver the anticipated risk mitigation.

## IEEE 802 CATEGORIES FOR NETWORKING STANDARDS

The IEEE networking standards continue to evolve to reflect technological and behavioral changes since the initial working group for IEEE Standards Project 802 meeting was convened in October 1979. There were significant expansions in the 2014 version to reflect the increasingly varied use of interconnected wireless technologies. Networks described by the IEEE 802

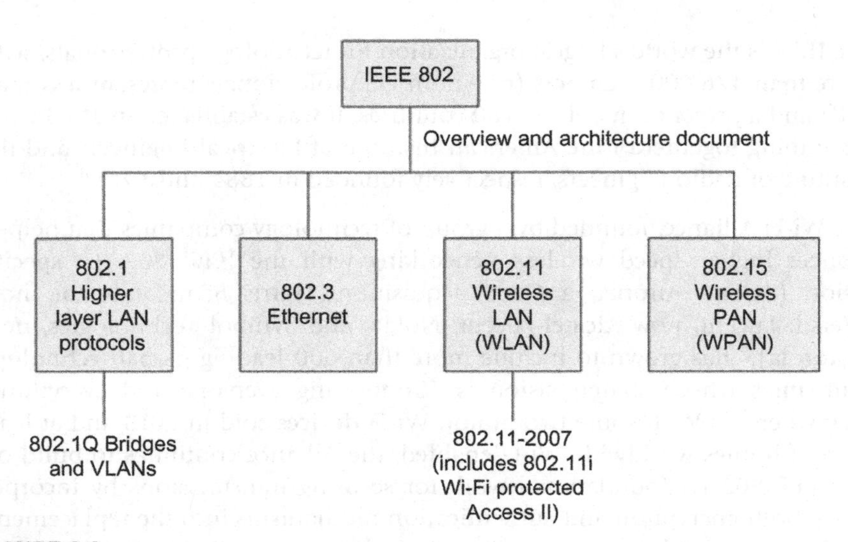


**FIGURE 1.1**
Where to Find Security Discussions in IEEE 802 Documents.

standards, from least to most geographically extensive, are the personal area network (PAN), local area network (LAN), wide area network (WAN), metropolitan area network (MAN), and regional area network (RAN).[12]

The wireless discussion covers those networks and devices that communicate over "free" space, rather than through fixed (e.g., optical fire, copper wire, cable) facilities. The wireless and wired networks do not exist in strictly parallel universes, of course. Signals travel over a variety of channels; wired and wireless worlds can converge in the course of a single communications transmission. Therein lies a key challenge for building resilient, secured, and assured communications: It is essential that the specific properties, including vulnerabilities, of each communication medium (copper, cable, radio, or optical) transmitting or receiving a specific signal be understood.

The reference model used by IEEE 802 is based on the lowest two layers of the familiar seven-layer OSI reference model, the physical and data link layers. The standard further divides the data link layer into the MAC and logical link control sublayers. Although creation of a technical advisory group with a focus on 802 security has been proposed, at present only four of the 802 family branch standards include discussions of security, as noted in Figure 1.1. Specific considerations for wireless networking functionality is captured in five IEEE 802 standards:

- 802.11—Wireless Local Area Network (WLAN or Wi-Fi)
- 802.15—Wireless Personal Area Network (WPAN; includes near-range technologies like Bluetooth under 802.15.1, ZigBee under

802.15.4, mesh networking under 802.15.15, and wireless body area network (WBAN, used in medical applications) under 802.15.6; optical spectrum connectivity is also possible over the infrared spectrum)
- 802.16—Broadband Wireless Access (Wireless Metropolitan Area Network or WMAN; also referred to as WiMAX)
- 802.20—Mobile Broadband Wireless (working group is currently in hibernation)
- 802.22—Cognitive Wireless Regional Area Network (WRAN; proposed to enable rural broadband wireless access to the frequency range from 54 to 862 MHz, the TV signal whitespace[13] and other unlicensed spectrum[14])

## WIRELESS LAN OR WI-FI (802.11X)

In the LAN articulation, the 802.11 protocol family displays different characteristics that are relevant to a variety of communications scenarios. These standards address wireless management functions: configuration, fault, performance, security, and accounting. The general characteristics of different 802.11 connectivity standards are indicated in Table 1.1, which captures hypothetical as well as typical data transmission rates achieved. These rates vary according to signal strength and are affected by distance traveled, environmental factors, frequency band noise, and service-level configurations (e.g., encryption, authentication, or authorization processes). Other standards have been defined for robust audio video streaming (802.11aa) and television white space (TVW) operation (802.11af).

Currently the most common new wireless network infrastructure deployments follow 802.11g and 802.11n, although legacy deployments may include 802a and 802b. The recently ratified (late 2013) 802.11ac standard has the potential for impressive throughput and is being built into mobile user devices, but its actual performance is dependent on the capacity of the wireless routers and access points to which devices are connected. Routing through devices that perform at a lower level degrades the hoped-for performance of higher-end devices. In addition to transmission and connectivity standards within the 802.11 standards family are security protocols like 802.11i and 802.11w-2009. The latter standard addresses prevention of incidents like denial of service (DoS) attacks through additional encryption security features. Information about the official IEEE 802.11 timeline for working groups assigned to specific projects is available through the IEEE website.[15]

**Table 1.1** Overview of Wireless Networking Standards Based on IEEE 802.11

| Standard | Ratified | Frequency Band | Max Data Rate (MBPS) | Typical Throughput (MBPS) | Max Nonoverlapping Channels | Comments |
|---|---|---|---|---|---|---|
| 802.11 | 1997 | | | | | |
| 802.11a | 1999 | 5 GHz | 54 | 25 | 24 (20 MHz)<br>12 (40 MHz) | More signal loss than at lower frequencies<br>Typical uses: larger corporate networks or WISP outdoor backbone networks |
| 802.11b | 1999 | 2.4 GHz ISM | 11 | 6.5 | 3 (20 MHz) | |
| 802.11g | 2003 | 2.4 GHz ISM | 54 | 8 (mixed b/g)<br>25 (only g) | 3 (20 MHz) | Most prevalent for residential and commercial |
| 802.11n | 2009 | 2.4 GHz<br>5 GHz | 248 | | For 2.4 GHz:<br>3 (20 Mhz)<br>1 (40 MHz; not recommended)<br>For 5 GHz:<br>24 (20 MHz)<br>12 (40 MHz) | Adds multiple input/multiple output and spatial multiplexing<br>Requires more electrical power for operation |
| 802.11ac | 2013 | 5 GHz | 1300 | 720 | 80 MHz<br>160 MHz<br>(plus 20 MHz and 40 MHz) | Typical uses: smartphones and TVs, laptops, desktops.<br>Built-in beam forming for decreased signal loss |
| 800.11ad (WiGig) | 2012 | 60 GHz (unlicensed radio spectrum ~57–66 GHz) | 7000 | 1540 | (detail needed) | Short-range uses<br>Built-in beam forming for decreased signal loss |

*Air802 web page. Retrieved from http://www.air802.com/files/802-11-WiFi-Wireless-Standards-and-Facts.pdf; Gordon Kelly, "802.11ac vs 802.11n WiFi: What's the difference?"* Forbes. *Retrieved from http://www.forbes.com/sites/gordonkelly/2014/12/30/802-11ac-vs-802-11n-wifi-whats-the-difference/; IEEE Standards Association. Retrieved from http://standards.ieee.org/about/get/802/802.11.html*

## WIRELESS MAN OR WIMAX (802.16)

Originally known as worldwide interoperability for microwave access, WiMAX is not now considered an acronym but, rather, the trademark for the private sector industry trade association, WiMAX Forum. The technology refers to large WMANs, usually managed by an Internet service provider (ISP), government entity, or business. Examples of a business delivering WiMAX services would be a hospital in a large, somewhat rural area that offers area businesses and residents broadband access over the infrastructure that it is building for connecting with medical offices throughout the region. WiMAX specifies two basic security services: authentication and confidentiality (viewing data messages is restricted to authorized devices). More robust security is enabled through additional security services not specified in the standard. In addition to susceptibility to the same security challenges of 802.11 and wired networks, WiMAX networks are also vulnerable to various disruptions between WiMAX nodes, whether deployed as a nonline-of-sight or line-of-sight system.[16]

## WIRELESS PAN (802.15)

WPANs are characterized by casual, ad hoc interconnections between devices typically in close range to one another (up to 33 feet for Class 2 devices like mobile devices and smart card readers and up to 3 feet for Class 3 devices like Bluetooth adapters for connecting a computer and keyboard or a mobile phone to a car's speaker). Class 1 devices—for example, some USB adapters and access points—can operate over distances of up to 328 feet or about 100 meters, however.

Bluetooth devices do not necessarily connect on a one-to-one basis. One interconnectivity model is a *piconet,* which is composed of two or more Bluetooth devices within close physical proximity, operating on the same channel, and using the same frequency hopping sequence, as in the example of the Class 3 devices above. Bluetooth operates in the same frequency band as 802.11b/g networks, a fairly crowded band. Unlike 802.11b/g networks, which have fixed frequencies, Bluetooth employs frequency hopping spread spectrum (FHSS) technology for transmissions. Although this will not resolve transmission security concerns significantly, using FHSS does reduce transmission errors and signal interference. Transmission power is also negotiated between Bluetooth devices using radio link power control, whereby the devices gauge received signal strength and request that another device adjust its radio power level up or down.

Another Bluetooth networking topology is the *scatternet,* a chain of piconets in which a Bluetooth device may be designated as *master* in one piconet, but as a *slave* in one or more other piconets simultaneously. A specific device can only be master of one piconet and devices must have point-to-multipoint capability to participate in a scatternet.[17] The dynamic topology created can change during a given session, depending on a device's location or relationship with respect to the master device.

Although the Bluetooth specifications define several security modes, security controls are initiated at different points in the challenge–response handshake process, thus leaving connections potentially vulnerable. As with any wireless networking technology, Bluetooth connections are vulnerable to a variety of threats including DoS, eavesdropping, man-in-the-middle (MITM), message corruption, resource misappropriation, and spoofing. Because end-to-end protection cannot be assured, organizations and individuals should follow a practice of "use as needed/use when needed" and change manufacturer default settings that enable Bluetooth in mobile devices.

Authentication, confidentiality, and authorization are the three basic security services specified in the Bluetooth standard. Devices operate in one of four security modes, with Security Mode 1 on NIST's do-not-use list because it offers no protection: A device in this mode is indiscriminate, although it will participate in security mechanisms if another device does the initiating. Security Mode 2 is enforced at the service level and allows for authorization; rules for access to some services and not others can be specified. In Security Mode 3, security procedures are initiated at the link layer, so prior to completion of the physical link, unlike Mode 2. Still, NIST recommends using service-level security as well to control for authentication abuse. Security Mode 4 uses Secure Simple Pairing (SSP) passing keys between devices, but because the security procedures initiate after the physical and logical links are established, and because Bluetooth v2.0 and earlier devices do not support the SSP feature, NIST recommends Mode 3.[18]

The Bluetooth Low Energy (BLE) guideline was released in 2010 in the Bluetooth v4.0 specification and, as its name indicates, is a power-saving technology that is especially useful in environments where availability is a high priority. It is also more expansive in terms of the number of slaves a master device can associate with in a piconet: unlimited, as opposed to the earlier versions' limitation of up to 255 inactive slaves. BLE does not, on the other hand, support scatternet topology. BLE is embedded in smartphones, laptops, medical devices, sensors, and other applications that benefit from the technology's key differentiators: lower power consumption, reduced memory requirements, efficient discovery and connection procedures, short packet lengths, and simple protocols and services.[19] Wi-Fi Direct has been touted as a competitor to BLE on the basis of both data exchange rate speeds and operating distances.

## RADIO FREQUENCY IDENTIFICATION (RFID)

Another significant wireless technology that now has implications for networking had its roots in World War II. Initially designed as a way of identifying aircraft as friend or foe, RFID is a form of automatic identification and data capture (AIDC) technology that uses radio wave spectrum to pass information between two objects: an identifying tag (like the inventory control devices used in retail stores or public libraries to reduce the likelihood of items leaving the premises undetected) and a reader (like the wands waved over ski passes in lift lines or an employee's building access badge). Tags may be passive (no internal power supply), active (internal power supply), semi-passive (internal power supply for circuitry or sensor support, but not communication) or semi-active (internal power supply but dormant, not communicating, until energized by a reader).

Because they can be miniaturized, RFID tags can also be used to track living objects; one use encouraged by the US Department of Agriculture was to record which livestock had received hormones or medications so that dosages could be controlled without human error. (A more recent use is as microchips in pets: Dog tags are so 20th century.) Other early uses were for tracking hazardous cargo (e.g., nuclear materials) and vehicle location and other information. The latter led indirectly to the development of automated toll payment systems by a group of enterprising scientists from Los Alamos in the mid-1980s.[20]

As various industries recognized the applicability of RFID technology platforms to automate processes to manage, control, and audit key business functional areas—inventory, warranty, fleet, warehousing, facilities access, payment processing, antitheft, for example—information about RFID-tagged objects was increasingly transmitted over the Internet. Research labs opened by the Auto-ID Center developed associated interface protocols, an identification scheme for data (Electronic Product Code or EPC system), and network architecture. The Uniform Code Council licensed this technology in 2003, then joined with European Article Numbering (EAN International) to promote the technology and standards for use, ultimately merging into Electronics Product Code Global Incorporated (EPCglobal). The other standards body for RFID is the ISO. Along with the International Electrotechnical Commission (IEC), it has issued several standards that address recommended development and deployment:

- ISO/IEC 15961 (data protocol: application interface)
- ISO/IEC 15962 (data protocol: data encoding rules and logical memory functions)
- ISO/IEC 15963 (unique identification for RFID tags)
- ISO/IEC 18000 series (parameters for air interface at different frequency levels)
- ISO/IEC 18046 series (device performance test methods)
- ISO/IEC 24791 series (software system infrastructure)
- ISO/IEC 18047 series (device conformance test methods)
- ISO/IEC 29167 (AIDC techniques).[21]

RFID technology deployments are a key part of the IoT landscape, often crossing organizational boundaries to streamline processes within a supply chain, for example. Negotiating agreements and implementing security mechanisms about how the links in that chain are managed by different organizational jurisdictions are challenges that must be coordinated by both the business and technical teams of those organizations.

## SUPERVISORY CONTROL AND DATA ACQUISITION

Supervisory Control and Data Acquisition (SCADA) and other related industrial control systems (ICS) technologies are embedded in many of our critical infrastructure sectors on the list of 16 published by the US Department of Homeland Security (DHS), for example, energy; dams; water and wastewater; nuclear reactors, materials, and waste; transportation (including air traffic and high-speed rail control); government facilities (especially prisons and correctional facilities); chemical; and critical manufacturing. These technologies support, automate, and control certain physical, mechanical, electrical, hydraulic, pneumatic, or distribution processes. Initially deployed as self- or facility-contained, independent systems, digital technology is increasingly replacing the manual or human component. This allows remote diagnostics and maintenance, in addition to concatenating distributed systems into a system of systems, as in smart power grids, buildings, and manufacturing.

This shift to dependence on digital technology, the insertion of IT capabilities to replace or supplement physical control mechanisms, is not seamless. Underlying expectations differ. For one, ICS legacy technical components are characterized by stability and have, typically, been deployed for decades; components like sensors, actuators, and controllers are not swapped out and upgraded in three-year cycles, but are "built to last." Operating systems for computers within the ICS may have exceeded manufacturer end-of-life and no longer be supported with patches. Likewise, replacement parts for infrastructure components can be difficult to source. And yet, the first priority for these systems is robustness: Availability, consistent uptime within tight latency and mean time between failure boundaries, is the primary concern to meet the objectives of real-time monitoring and response. Any software changes must be fully tested under operational environment conditions, but rollouts cannot have a negative impact on the actual operational performance. System outages should be scheduled weeks in advance. In addition, increased interdependency of system elements must be balanced with concerns about cascading failures. Any changes in one part of the system that could affect another must be anticipated and mitigating responses prepared. The 2003 Northeast US power failure that affected 50 million people, although not a consequence of a cyberattack, is just one reminder of the butterfly wing vulnerability in the critical infrastructure arena.

Distributed SCADA systems, such as those in an electrical distribution scenario, gather information from remote sensors in the field and send it to a centralized facility where a human operator observes the information in textual or graphic format. Such SCADA systems are designed for fault-tolerance. According to the specific implementation, communications may use a variety of different telemetry media: copper, cable, fiber, or radio frequency (e.g., broadcast, microwave, or satellite). Communications topologies also vary and can include point-to-point (simplest but most expensive because of the number of individual channels required), series, series-star, and multi-drop. The series configuration reduces cost, but the channel-sharing approach has a negative impact on SCADA operation efficiency and system complexity, as does the one-channel-per-device configuration of series-star and multi-drop topologies.[22]

Concern about the resiliency of SCADA and other ICS implementations in critical infrastructure sectors has increased with the latter's obvious vulnerability to a variety of cyberattacks. The energy sector, for example, experienced 41% of the 198 incidents addressed by DHS' ICS cyber response team in 2012. Of the 200 executives from critical utility enterprises in 14 countries who responded to a survey conducted in late 2010, 85% reported that they had experienced a network intrusion and 80% said they had experienced a large-scale DoS attack.[23] Understanding how to select and implement security control mechanisms and policies at all levels of these complex, increasingly Internet-based, and wirelessly connected systems will benefit the critical infrastructure enterprises themselves and all organizations, government, and individuals that depend on them for completing even the most mundane tasks like boiling water for a cup of tea or turning on lights in a critical care unit.

## WHERE DO WE GO FROM HERE?

As can be seen from the above discussion, there are a lot of moving parts to consider in how wireless devices talk and connect to one another. Almost any wireless devices can be construed as a wireless access point, with an accompanying set of risks. The operating environment for devices, the data that they carry or transmit, figures largely in whether and how those risks should be addressed. With mobile devices the operating environment is less certain than for fixed devices; a database server, by contrast, is located in a highly controlled operating environment. The security exposure of mobile devices is more situationally dependent—but can also be configured by the astute user when appropriate. Risks can be grouped into commonly encountered scenarios and addressed in a straightforward manner.

The enthusiastic adoption of wireless communications by individuals, corporations, organizations, educational institutions, and governments lends credence to the suggestion floated among IT professionals that the OSI model should be modified to include end users as Layer 8. By understanding trends in who is using wireless devices, the purposes behind their use, and the locations from which/to which communications are sent, specific techniques for hacking—or for shielding from hacking—can be proposed.

## ENDNOTES


1. Gartner, Inc. (11 November 2014), "Gartner Says 4.9 Billion Connected 'Things' Will Be in Use in 2015," Press release retrieved from http://www.gartner.com/newsroom/id/2905717.
2. Juniper Research (28 July 2015), "'Internet of Things' Connected Devices to Almost Triple to Over 38 Billion Units by 2020," Press release retrieved from http://www.juniperresearch.com/press/press-releases/iot-connected-devices-to-triple-to-38-bn-by-2020.
3. DHL and Cisco (15 April 2015), "Internet of Things in Logistics Trend Report," retrieved from http://www.dhl.com/en/about_us/logistics_insights/dhl_trend_research/internet_of_things.html#.ViXRnqQqpiI.
4. International Telecommunication Union (2009), "Security in Telecommunications and Information Technology: An Overview of Issues and the Deployment of Existing ITU-T Recommendations for Secure Telecommunications," retrieved from http://www.itu.int/dms_pub/itu-t/opb/hdb/T-HDB-SEC.04-2009-PDF-E.pdf.
5. NIST (13 September 2011), "U.S., Europe Collaborating on Smart Grid Standards Development," Press release. Retrieved from http://www.nist.gov/smartgrid/grid-091311.cfm.
6. http://www.3gpp.org/about-3gpp.
7. https://www.ietf.org/about/.
8. Retrieved from the IEEE website http://www.ieee.org/about/today/index.html.
9. "Wi-Fi Alliance," retrieved from https://en.wikipedia.org/wiki/Wi-Fi_Alliance.
10. Wi-Fi Alliance. Web page "Who we are," retrieved from http://www.wi-fi.org/who-we-are.
11. Based on slides summarizing findings from John L. Hennessy and David A. Patterson (2009), "Computer Architecture: A Quantitative Approach," retrieved from http://www.cs.columbia.edu/~sedwards/classes/2012/3827-spring/advanced-arch-2011.pdf. Ken Polsson posts his "Chronology of Microprocessors" from 1958 to 2015 online at http://processortimeline.info/.
12. Excellent visuals showing the relationship among these categories can be retrieved from http://www.rfidc.com/docs/introductiontowireless_standards.htm.
13. Apurva N. Mody (June 2010) slide presentation. Retrieved from http://www.ieee802.org/22/Technology/22-10-0073-03-0000-802-22-overview-and-core-technologies.pdf.
14. The FCC ruled in early August 2015 on opening up channels in the TV and 600 MHz bands. Some implications for urban areas are discussed by Erika Morphy (14 August 2015), "How New 'White Space' Rules Could Lead to an Urban Super-Wi-Fi," *Computer World*. Retrieved from http://www.computerworld.com/article/2970867/wireless-networking/how-new-white-space-rules-could-lead-to-an-urban-super-wi-fi.html.
15. http://grouper.ieee.org/groups/802/11/Reports/802.11_Timelines.htm.
16. Shirley Radack (December 2010), "Securing WiMAX Wireless Communications," NIST Bulletin. Retrieved from http://csrc.nist.gov/publications/nistbul/december2010-bulletin.pdf.
17. Ericsson (June 2004), "Scatternet—Part 1, Baseband vs Host Stack Implementation," retrieved from https://www.bluetooth.org.
18. NIST, Guide to Bluetooth Security.
19. John Padgette, Karen Scarfone, and Lily Chen (June 2012), "Guide to Bluetooth Security," National Institute of Standards and Technology SP 800-121 Rev. 1, p. 5. Retrieved from http://csrc.nist.gov/publications/drafts/800-121r1/Draft-SP800-121_Rev1.pdf.
20. Mark Roberti (16 January 2005), "The History of RFID Technology," *RFID Journal*. Retrieved from http://www.rfidjournal.com/articles/view?1338/.
21. French National RFID Center list of published and in-process standards. Retrieved from http://www.centrenational-rfid.com/list-of-published-isoiec-standards-article-106-gb-ruid-202.html.

22. Keith Stouffer, Victoria Pillitteri, Suzanne Lightman, Marshall Abrams, and Adam Hahn (May 2015; pre-publication), "Guide to Industrial Controls Systems," NIST SP 800-82 Rev. 2. Retrieved from http://csrc.nist.gov/publications/nistpubs/800-82r2/sp800-82_r2_PRE-PUBLICATION.pdf.
23. Stewart Baker, Natalia Filipiak, and Katrina Timlin (2011), "In the Dark: Crucial Industries Confront Cyberattacks," McAfee and Center for Strategic and International Studies. Retrieved from http://www.mcafee.com/us/resources/reports/rp-critical-infrastructure-protection.pdf.

CHAPTER 2

# Wireless Adoption

## WIRELESS INNOVATION AND ADOPTION: EARLY MARKET PENETRATION

The use of wireless communications technology based on radio wave spectrum has exploded in terms of number of transmissions, frequencies licensed, user demographics, and use case scenarios since Marconi's first experiments in 1895. And while the profile for adoption of wireless technology follows the familiar bell curve described by Everett M. Rogers in his *Diffusion of Innovations* (first published in 1962), those who represent the five different types of technology adopters—innovators (2.5%), early adopters (13.5%), early majority (34%), late majority (34%), and laggards (16%)[1]—have shifted.[2]

Although the innovators were often individual inventors or associated with research labs (with some degree of government funding), the early adopters in the beginning of the 20th century tended to be involved in transportation (including aviation and maritime activities, e.g., open-sea radio signaling during the *Titanic* disaster in 1912, an improvement over the use of carrier pigeons and visual signaling with flags),[3] law enforcement, national defense (e.g., in the Anglo-Boer War, 1899–1902), utilities (e.g., telecommunications and energy), and news media. Individual early adopters outside those sectors were more likely to be hobbyists who scanned *Popular Mechanics* for technological updates and how-to instructions. Mass marketing campaigns to promote the adoption of technological changes appeared much later, as did adoption by retail and manufacturing businesses and civilian government agencies.

The first AM radio broadcast for music and entertainment was in 1906. The Radio Act of 1912 mandated that amateur (ham) radio operators be licensed and that ships deploy 24-hour radio service with a trained operator (in part a response to the inadequate rescue response to the *Titanic*). The American Radio Relay League, founded in 1914, helped organize relay stations for more efficient and reliable long-distance signaling[4] similar to the way routing tables organize telephony transmissions. The Federal Communications



Hacking Wireless Access Points. DOI: http://dx.doi.org/10.1016/B978-0-12-805315-7.00002-4

Commission (FCC) was established by the Communications Act of 1934, in part to address inadequacies in the earlier Radio Act of 1912 and subsequent Radio Act of 1927, under which the Federal Radio Commission was formed to address broadcasting.[5]

In 1933, the Bayonne, New Jersey, police department launched two-way AM radio use in patrol cars, five years after Detroit police first started using regular one-way radio communications in patrol cars. The two-way radios combined transmitter and receiver. In 1946, a taxi driver in St. Louis, Missouri, made the first mobile telephone call over a Southwestern Bell service.[6] This was the same year that newspaper cartoon detective, Dick Tracy, started using his two-way radio watch (the two-way TV watch didn't show up until 1964) to communicate with others on the police force in his pursuit of bad actors.[7] By 1948, wireless telephone service was available in 100 cities, although the service was very limited. Essentially a party line over which calls were placed by an operator, the service could handle only three subscriber calls simultaneously. The metropolitan area's infrastructure consisted of a single transmitter, centrally located, signaling over the limited radio spectrum licensed by the FCC for this use. The equipment in the trunk required to run this early mobile telephone, which weighed in at about 80 lb and dimmed the headlights when used, would make the 1983 Motorola "brick" cellular phone—released 35 years later and weighing in at a mere 5 pounds—feel more truly mobile.[8] Monthly subscriptions and local call charges were expensive in 1948: the equivalent of almost $200 today. The service was typically used by businesses: utility companies, truck fleet operations, and journalists. Adoption was limited to about 5000 customers making some 30,000 calls per week.

## WIRELESS MARKET PENETRATION: US CONTEXT

By contrast, in 2014, 90% of American adults owned a cell phone,[9] about 221.4 million individuals,[10] and collectively used about 204.6 billion monthly voice minutes in 2014.[11] Similar to IBM founder Thomas J. Watson's underestimation of the potential market appeal of computers, a 1983 McKinsey & Co. report prepared for AT&T predicted 900,000 US cellular subscribers by 2000. That milestone was reached in 1987, however; there were 109 million subscribers in the United States by 2000.[12] This shift to deep penetration of the marketplace took time and development of contributing factors: battery technology to extend time between charges; invention of transistors (1947) to replace large, heavy, environmentally fragile vacuum tubes and enable a broad range of convenient consumer products from hearing aids to pocket-sized GPS to nanoscale devices;[13] expanded spectrum availability and accessibility through a well-defined system for its allocation; favorable economic and cultural climates that championed technological possibility and perceived value; development of vernacular, human readable computer programming code; and even marketing approaches that made exotic innovation somehow familiar and affordable.

It is hard to imagine being completely removed from the reach of radio spectrum technology. Even those who venture into austere or isolated environments can find hardened personal devices like avalanche transceivers and others for location and communication services in "no service" areas. Wildlife has been tracked via radio since the late 1950s. We are in an unusual time in which communications technology is so embedded into our environment that we become de facto users, whether because of the credit cards or passport we carry, our fitness sensors, our mood-sensing shoes,[14] or our baby's rubber ducky, Edwin.[15]

There were some 304,360 cell sites in the United States as of year-end 2013, which represent an 87% increase over a period of 10 years (December 31, 2003, through December 2013).[16] Although the FCC expects the demand driving that growth to slow, anticipated qualitative changes to the traffic (e.g., increased video streaming, which is more data-intensive) will require continued investment from telecom service providers in mobile backhaul technology—as much as $43 billion, collectively, between 2013 and 2017—to connect with mobile switching centers. From these centers, connections are made to "the provider's core network, the public switched telephone network (PTSN), or the Internet" to ensure routing and onward transmissions of traffic.[17]

## WIRELESS PENETRATION RATES: GLOBAL CONTEXT

The key consensus point around why we use wireless communications in the United States, whether data or voice, is that it is convenient. We are willing to tolerate performance failures that would be unacceptable from a wireline service provider with respect to level of service expectations. Indeed, the dropped calls, lack of vocal clarity, and battery insufficiency have provided us a new set of socially acceptable excuses for not wanting to communicate with a caller. The inefficiencies are preferable to being tethered to a landline, even one with a portable handset. Mobile carriers continue to add services like conference calling to mimic the features available on fixed lines. In addition, even as the uptake in wireless subscriptions continues, the cost of use is going down and compares favorably to wireline use, as shown in Fig. 2.1.

For other parts of the world, the initial game changer that wireless service introduced was building up capacity for communications. The plain old

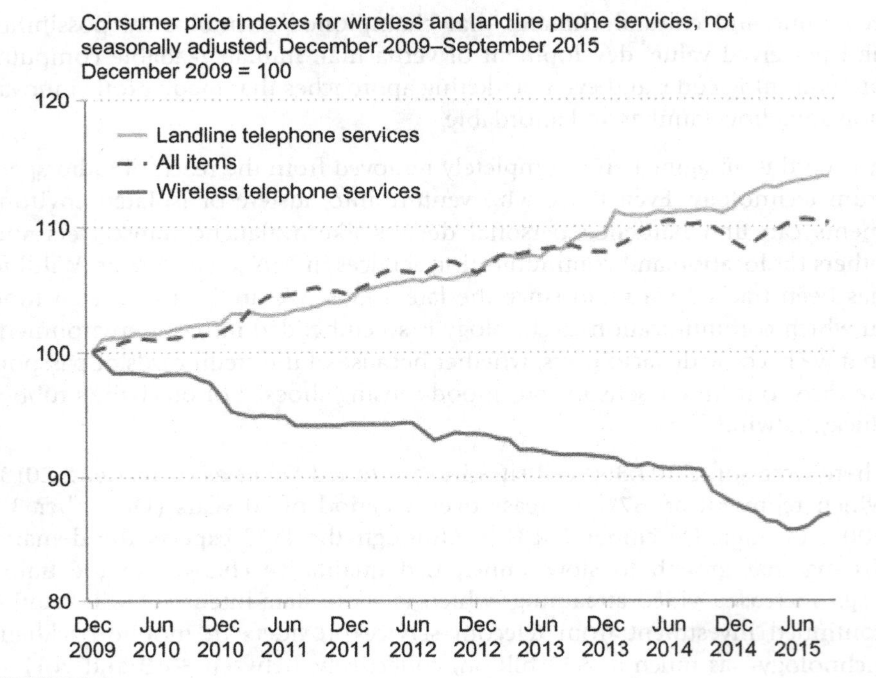


**FIGURE 2.1**
Consumer price index comparison of telephone services.[18]

telephone system (POTS) infrastructure for voice (and data) services was not as dense elsewhere as in the United States, even in what are now considered Eurozone nations. As a case in point, the average wait time for obtaining landline service in some Central and Eastern European countries (CEEs) in 1990 was 11.5 years.[19] Data collected by the ITU for 2013 show significant mobile phone penetration, with the percentage uptake by individual subscribers ranging from 81.5 (Bulgaria) to 96 (Czech Republic).[20]

The increase in penetration rates between 2005 and 2015 for mobile phone subscribers in Africa is significant and exceeds fixed-telephone subscriptions dramatically. Table 2.1 shows comparative data gathered by the ITU for penetration rates in UN-defined global regions. Clearly, individuals in Africa and the Arab States have largely elected to skip the POTS generation of technology in favor of adopting newer wireless technology that is less costly to deploy over large areas. Across Sub-Saharan Africa, the aggregated penetration rate for mobile broadband subscriptions grew 156% between 2010 and 2011, slowed to 85% between 2011 and 2012, then to 21% between 2012 and 2013, to 25% between 2013 and 2014, and is estimated to have grown 35% between 2014 and 2015. Over this same five-year time period, the fixed

**Table 2.1** Longitudinal Comparison of Technology Penetration Rates by UN Region[a]

| | 2005 | 2006 | 2007 | 2008 | 2009 | 2010 | 2011 | 2012 | 2013 | 2014 | 2015[b] |
|---|---|---|---|---|---|---|---|---|---|---|---|
| ***Fixed Telephone Subscriptions*** | | | | | | | | | | | |
| Africa | 1.5 | 1.5 | 1.5 | 1.5 | 1.6 | 1.5 | 1.4 | 1.3 | 1.1 | 1.2 | 1.2 |
| Arab States | 9.4 | 9.6 | 10.1 | 10.3 | 9.9 | 9.8 | 9.8 | 9.6 | 8.9 | 8.1 | 7.3 |
| Asia and Pacific | 15.1 | 15.5 | 15.3 | 14.9 | 14.9 | 14.2 | 13.7 | 13.3 | 12.5 | 11.9 | 11.3 |
| CIS | 23.0 | 24.7 | 25.8 | 26.0 | 26.1 | 26.2 | 26.1 | 25.7 | 24.6 | 23.9 | 23.1 |
| Europe | 45.5 | 45.3 | 43.7 | 42.7 | 43.6 | 42.8 | 41.6 | 40.3 | 39.4 | 38.3 | 37.3 |
| The Americas | 33.0 | 31.9 | 30.9 | 31.2 | 29.9 | 29.2 | 28.3 | 27.6 | 27.0 | 26.2 | 25.4 |
| ***Mobile-Cellular Telephone Subscriptions*** | | | | | | | | | | | |
| Africa | 12.4 | 17.8 | 23.5 | 32.2 | 38.0 | 45.4 | 52.3 | 58.9 | 65.6 | 71.2 | 73.5 |
| Arab States | 26.8 | 38.8 | 52.6 | 63.2 | 76.5 | 87.9 | 99.1 | 105.4 | 110.4 | 109.7 | 108.2 |
| Asia and Pacific | 22.6 | 28.8 | 37.1 | 46.6 | 56.3 | 67.3 | 76.5 | 80.9 | 86.7 | 90.6 | 91.6 |
| CIS | 59.7 | 81.8 | 96.1 | 111.6 | 126.8 | 134.2 | 127.2 | 130.5 | 137.0 | 137.7 | 138.1 |
| Europe | 91.7 | 101.2 | 111.7 | 117.0 | 116.8 | 115.0 | 117.9 | 119.6 | 120.1 | 120.5 | 120.6 |
| The Americas | 52.1 | 62.0 | 72.1 | 80.8 | 87.3 | 94.0 | 100.6 | 103.9 | 107.7 | 108.2 | 108.1 |
| ***Active Mobile-Broadband Subscriptions*** | | | | | | | | | | | |
| Africa | N/A | N/A | N/A | N/A | N/A | 1.8 | 4.6 | 8.5 | 10.3 | 12.9 | 17.4 |
| Arab States | N/A | N/A | N/A | N/A | N/A | 5.1 | 13.1 | 16.1 | 27.3 | 36.1 | 40.6 |
| Asia and Pacific | N/A | N/A | N/A | N/A | N/A | 7.4 | 11.0 | 15.3 | 18.5 | 29.7 | 42.3 |
| CIS | N/A | N/A | N/A | N/A | N/A | 22.0 | 31.3 | 35.3 | 42.3 | 46.9 | 49.7 |
| Europe | N/A | N/A | N/A | N/A | N/A | 30.5 | 39.4 | 49.1 | 56.1 | 69.3 | 78.2 |
| The Americas | N/A | N/A | N/A | N/A | N/A | 24.6 | 34.1 | 41.9 | 55.7 | 67.3 | 77.6 |
| ***Fixed Broadband Subscriptions*** | | | | | | | | | | | |
| Africa | 0.0 | 0.1 | 0.1 | 0.1 | 0.1 | 0.2 | 0.2 | 0.2 | 0.3 | 0.4 | 0.5 |
| Arab States | 0.3 | 0.5 | 0.9 | 1.3 | 1.6 | 1.9 | 2.2 | 2.6 | 3.2 | 3.4 | 3.7 |
| Asia and Pacific | 2.2 | 2.8 | 3.2 | 4.0 | 4.7 | 5.5 | 6.4 | 7.0 | 7.8 | 8.3 | 8.9 |
| CIS | 0.6 | 1.3 | 2.3 | 4.3 | 6.1 | 8.0 | 9.2 | 11.0 | 12.7 | 13.1 | 13.6 |
| Europe | 10.9 | 14.8 | 18.4 | 20.4 | 22.1 | 23.6 | 24.8 | 25.7 | 27.7 | 28.6 | 29.6 |
| The Americas | 7.5 | 9.0 | 10.9 | 12.3 | 13.0 | 14.0 | 15.0 | 15.8 | 17.0 | 17.4 | 18.0 |

[a]*U.S. Bureau of Labor Statistics (28 October 2015), "Price Trends for Wireless and Landline Phone Services," December 2009–September 2015. Retrieved from http://www.bls.gov/opub/ted/2015/price-trends-for-wireless-and-landline-phone-services-december-2009-september-2015.htm.*
[b]*Estimate.*

broadband penetration rate has only grown minimally, and from a much lower starting base penetration rate.

In Europe and Asia, one can see a steady decline in fixed-line subscriptions over time. This phenomenon has also been noted in the United States among households. Statistics from the World Bank show that fixed telephone subscriptions have declined between 2011 and 2015 in 66% of countries worldwide, stayed fairly stable in 12%, and increased in 23% of the 206 countries worldwide for which data were reported.[21]

## WIRELESS PENETRATION RATES: SECURITY RISK CONTEXT

By looking at both wireline and wireless penetration rates, one can begin to appreciate the relative vulnerability, in terms of communications capacity and resiliency, of different populations. The United States exemplifies an over-build condition, in which legacy wireline capacity is now layered with wireless capacity. Other countries, especially those that skipped over the POTS generation of technology, rely more heavily on their wireless infrastructure; out-of-band capability to support more robust authentication services, for example, is limited. Protecting that wireless infrastructure becomes more critical when it is, in effect, the only telecommunications infrastructure.

The expanded subscriber base for wireless communications worldwide, combined with the extension of wireless use case scenarios across all industries, makes WAPs more attractive targets to opportunistic bad actors. There is more potential gain. At the same time, tools for mobile hacking are becoming more prevalent, thus lowering the learning threshold (and cost) to those potential bad actors who are looking to reap benefits or wreak havoc.

At the 2013 Black Hat Conference, for example, versatile, economical tools and techniques publicized for maliciously targeting WAPs included the following functionalities:

- Transform another person's mobile phone into an audio or video bug, operated by a command and control (C&C) web server
- Extrude information from pacemakers and other embedded medical devices
- Exploit Android operating system "master key" vulnerability by hiding malicious code behind trusted, cryptographically vetted signatures
- Install surveillance tools that bypass malware detection and mobile device management controls, and then gather text and email location information
- Intercept voice, data, and text traffic, as well as clone connected devices
- Build malicious chargers for iPhones
- Implement a surreptitious, large-scale sensor-based tracking system that records activities of groups and individuals
- Manipulate Flash storage memory to control devices and even ICS by hiding malicious files or compromising performance

- Decrypt traffic sent by Bluetooth-enabled "smart" devices
- Clone a RFID tag (or access badge) by using a microcontroller to modify the reader[22]

The above tools and techniques are already common knowledge. Innovation in hacking, cracking (defeating security controls in wireless systems), and skyjacking (mobile hostage taking) WAPs continues enthusiastically, as does malicious interest in elevating privileges or gaining unauthorized access to mobile devices on various platforms: jailbreaking (iOS), rooting (Android), and unlocking (Windows).[23]

The need for securing mobile devices is a recurring theme among researchers and IT professionals as the number of devices, users, and applications—as well as the diversity of use cases—continues to proliferate. Concerns about the compromise of a single WAP that can then create an opening into a trusted network environment are captured in "top security threat" discussions. These threats act at multiple levels of the OSI stack and can be used as part of a layered attack against a government agency, public utility, or corporation, as well as against individuals. The potential (and realized) threats against WAPs often mirror those against corporate data centers. These threats include the following attack strategies.

## Attack Strategy: Mobile Payment

Attacks aimed at acquiring credentials for accessing bank accounts, social security numbers, and credit cards make headlines annually. One attack approach is extracting information stored in bulk on cloud servers. A predictable adaptation, as centralized servers are more tightly protected and mobile payment applications are adopted, is malware aimed at gathering payment credentials from multitudes of devices.[24]

## Attack Strategy: Malware Transmission

In one articulation, watering hole techniques attract unwitting visitors to legitimate news sites that are contaminated with hidden malware. Compromised devices can then pollute the devices belonging to colleagues and friends, even those who only respond to messages from trusted contacts. Spreading malware from mobile device to mobile device across a trusted network of other wireless users is just one attack scenario if AV controls are not installed, are managed ineffectively or not at all, or if the malware has not already been identified.[25] Zero-day attacks and the proliferation of polymorphing malware (malware code that constantly changes, making signature-based detection difficult) compromise the effectiveness of AV mechanisms.

## Attack Strategy: C&C (ICS Environment)

Another attack scenario is infection of a wired ICS network by a wireless device if AV controls are inadequate at the device or server level. One of the pieces of malware especially culpable is Havex, used to extrude data from ICS, especially those deployed in energy critical infrastructure that use Open Platform Communications or OPC. Havex is a remote access tool (RAT) type of malware. It leverages the functionality of the OPC, which consolidates information from numerous subsystems that operate independently.[26]

## Attack Strategy: DoS

In addition to the Havex OPC vulnerability, reports point to DoS risks in manufacturing industry production environments that rely on assembly-line processes, enterprise resource planning (ERP) systems, and building management solutions.[27]

## Attack Strategy: Suicide Malware

The next evolution of scareware and ransomware—the latter being malware that blocks even authorized access to data contained on a device and locks down the target's hard drive—could be malicious code that self-destructs. It is like a 21st century echo of the self-igniting magnetic tape through which the *Mission Impossible* TV team of the 1960s learned about the next mission, but the instructions this code destroys are the operating system instructions and memory. In the process of erasing evidentiary traces of the exploit, thus sabotaging incident response routines, this blastware will wipe out, not just lock out, all data on the device's hard drive if the code is modified in any way. This pernicious code can disrupt efforts to monitor possible advanced persistent threat (APT) activity.[28] It could also make holding a device hostage pending payment of extortion seem much the lesser of this new evil. By eliminating telltale traces of the attack pattern, it also makes reverse engineering the attack for the purpose of building proactive detection capability—and sharing alerts to other system operators—very difficult.

## Attack Strategy: SMS Infection Vector

Changes in user behavior make SMS an interesting attack vector. Smartphones are used five times more for texting in the United States, for example, than for phone calls.[29] This increased opportunity coincides with SMS use to spread and propagate malware (e.g., Trojans and worms) and SMS agents, through contact information stored on the initially compromised phone. Scams perpetrated through links buried in messages trigger premium services and SMS subscriptions. Fig. 2.2 shows top categories of SMS spam observed in 2014. To some analysts, this adaptation of old technology points

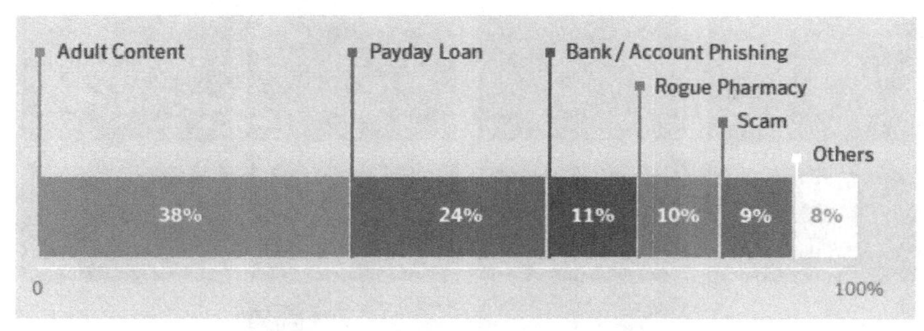


**FIGURE 2.2**
Top categories of observed SMS spam, 2014.[31]

to more unified communications and the need for user awareness "that threats can be delivered across a variety of areas.[30]

## Attack Strategy: Mobile Commerce

Online retailing makes good business sense from the perspective of businesses that can save dramatically by moving some transactions away from brick-and-mortar storefronts. Increasingly, consumers use multiple devices to accomplish purchases and often are even less prudent and security-conscious with mobile than with desktop devices.[32]

In addition to shopping online from mobile devices, consumers are paying online: 30% of ecommerce transactions are completed from smartphones.[33] A 2015 survey of 250 large organizations that are heavily engaged in online commerce (average revenue of respondents was more than $2.5 billion) indicated that between 25% and 49% ($92.3 million per year) of fraud incidents were attributed to transactions using mobile devices.[34]

## Attack Strategy: IoT

Onboard vehicle controls, embedded medical devices, smart clothing and household appliances, and ICS are just some of the use cases that comprise the IoT (or as some have suggested, the Internet of Threats). The potential for attacking these WAPs exists, as illustrated in the well-publicized Jeep hack demonstration in 2015.[35] Manufacturers and consumers need better guidelines about what kind of information is collected and transmitted from these devices and how such communications can be managed and protected.

The IoT also turns homes into computer centers, with a router/firewall "protecting" a zone containing home security systems, air conditioning/heating systems, washers, dryers, refrigerators, ovens, and other appliances. Homeowners are

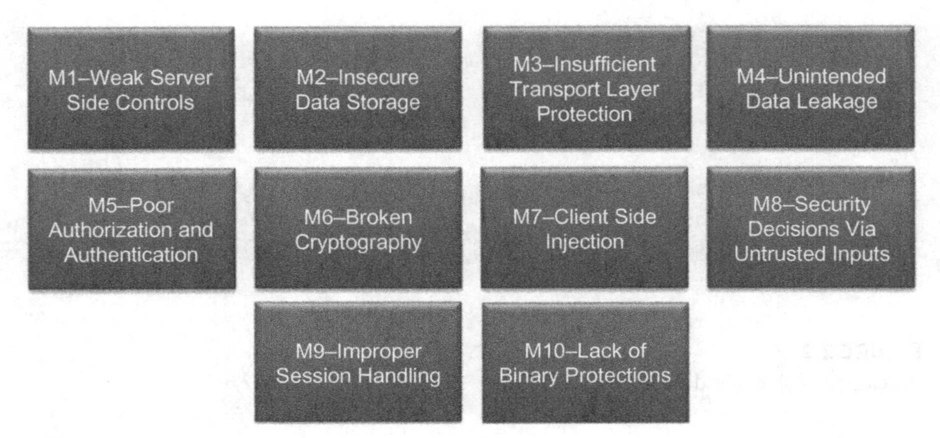


**FIGURE 2.3**
OWASP mobile top 10 risks.[38]

ill-prepared to defend against hackers who can apply decades of experience with attacking corporate and agency data centers. In fact, attacks on home routers were observed in 2014.[36]

### Attack Strategy: Layered Vulnerabilities

Of the 6.3 million Android apps analyzed by Symantec in 2014, 17% were characterized as disguised malware.[37] Vulnerabilities exist for Windows and iOS apps as well, in part because of vulnerability issues at various layers. The Open Web Application Strategy Project (OWASP) has identified 10 top mobile risks, as shown in Fig. 2.3.

## HACKING GOALS: STRATEGIES AND STEPS

In Chapter 3, Blurred Edges: Fixed and Mobile Wireless Access Points, you will learn about the various goals (end games) of hackers. These goals can be as simple as taking over a mobile device at a hackers' convention or as complex as transferring funds from a company's bank account into a relay bank account. The more complex the goal, the more likely the hacker will need to execute multiple hack techniques to overcome the defense in depth countermeasures protecting the assets desired by the hacker.

To achieve their goals, hackers must lay out a plan for their attack. The plan may be created by working backwards from their goals, for example, to obtain user identifier and password for a company's account, identify a company's personnel with needed credentials, identify the most likely person from whom to obtain credentials, identify locations where a targeted person communicates via wireless access, identify means to hack into these WAPs, and hack into these points when the person is present. Each part of the planned hack may use a different attack strategy or approach.

Each part of the attack strategy will typically follow the following steps:

- Reconnaissance—understanding the defenses protecting the assets associated with the hacker's goal. Reconnaissance can be either active (i.e., probing the network or computer) or passive (i.e., network sniffing).
- Scanning—using the information obtained during reconnaissance to further examine the assets' defenses and weaknesses.
- Gaining Access—exploiting the weaknesses identified during reconnaissance and scanning to gain access to the assets.
- Maintaining Access—keeping access to the assets for future exploitation and attacks.
- Covering Tracks—avoiding detection by the assets' countermeasures and removing evidence of the hack.

Discussions of successful hacks start in Chapter 4, Hacks Against Individuals. Although these discussions start with the hacker gaining access to an individual user's information assets, the reader should recognize that planning, reconnaissance, and scanning were required to achieve this access.

The enthusiasm to embrace wireless possibilities and explore a myriad of use case scenarios for consumer and organizational products and services has outpaced cautious discussion among the general public about the long-term impact of too much interconnectedness. In the United States, legal constraints on collecting and protecting ephemeral data, such as movement from one location to another, have been minimal.

Thus our stage is set for a deeper discussion about the differentiation between fixed and mobile WAPs, implications for their potential compromise, and interfaces with wired systems.

## ENDNOTES

1. E. M. Rogers (2003), Diffusion of Innovations (5th Edition), New York, NY: Free Press.
2. Rogers notes the evolution of his initial 1962 theory articulation to now include:
   - "Critical mass, defined as the point at which enough individuals have adopted an innovation that further diffusion *becomes* self-sustaining.
   - A focus on networks as a means of gaining further understanding of how a new idea spreads through interpersonal channels.
   - Re-invention, the process through which an innovation is changed by its adopters during the diffusion process."

   E. M. Rogers (2004), "A Prospective and Retrospective Look at the Diffusion Model," *Journal of Health Communications*, Vol. 9, 13–19. Retrieved from http://web.b.ebscohost.com.dml.regis.edu/ehost/pdfviewer/pdfviewer?vid=3&sid=b454b504-b2e0-4c05-9b92-f6cbb6ba0eae%40sessionmgr112&hid=101.

3. Federal Communications Commission (FCC) (Winter 2003–2004), "A Short History of Radio." Retrieved from https://transition.fcc.gov/omd/history/radio/documents/short_history.pdf.
4. American Radio Relay League, "*Ham* Radio History." Retrieved from http://www.arrl.org/ham-radio-history.
5. Jonathan S. Adelstein (7 March 2006), "Learning from History: Looking Back to Look Forward." Remarks at the National Emergency Number Association. Retrieved from https://apps.fcc.gov/edocs_public/attachmatch/DOC-264225A1.pdf.
6. Engineering and Technology History Wiki, "The Foundations of Mobile and Cellular Telephony." Retrieved from http://ethw.org/The_Foundations_of_Mobile_and_Cellular_Telephony.
7. Erin Blakemore (9 March 2015), "How Dick Tracy Invented the Smart Watch," *Smithsonian Magazine.* Retrieved from http://www.smithsonianmag.com/smart-news/how-dick-tracy-invented-smartwatch-180954506/?no-ist.
8. The brick was somewhat familiar to those who grew up watching comic TV-spy Maxwell Smart handle his shoe telephone in the 1960s.
9. Pew Research Center (January 2014), "Mobile Technology Fact Sheet." Retrieved from http://www.pewinternet.org/fact-sheets/mobile-technology-fact-sheet/.
10. Based on the 2014 US Census Bureau population estimate for people over 18 years of age, multiplied by 90%. CTIA estimates a total 315.9 million wireless subscribers for 2011. US Census Bureau figures retrieved from quickfacts.census.gov/qfd/states/00000.html. CTIA estimates retrieved from http://www.ctia.org/your-wireless-life/how-wireless-works/wireless-history-timeline.
11. CTIA—The Wireless Association (2015), "Year-End U.S. Figures from CTIA's Annual Survey Report." Retrieved from http://www.ctia.org/your-wireless-life/how-wireless-works/annual-wireless-industry-survey.
12. Engineering and Technology History Wiki, "The Foundations of Mobile and Cellular Telephony." Retrieved from http://ethw.org/The_Foundations_of_Mobile_and_Cellular_Telephony.
13. Sharon Gaudin (12 December 2007), "The Transistor: The Most Important Invention of the 20th Century?" *Computerworld.* Retrieved from http://www.computerworld.com/article/2538123/computer-processors/the-transistor--the-most-important-invention-of-the-20th-century-.html?page=2.
14. The 21st century answer to the mood ring: training shoes with screens that display their wearer's mood. Seriously. Sophie Curtis (16 November 2015), "Lenovo Unveils 'Smart Shoes' and Dual-Screen Smartwatch," *The Telegraph.* Retrieved from http://www.telegraph.co.uk/technology/news/11635136/Lenovo-unveils-smart-shoes-and-dual-screen-smartwatch.html.
15. Edwin is a recently released, $100 bathtub friend, night light, and Bluetooth-enabled sensor that can determine whether bath water or a baby's skin is the desired temperature. Wilson Rothman (12 November 2015), "Bluetooth Ducky, You're the One: A Toy for More Bath-Time Fun," *The Wall Street Journal.* Retrieved from http://www.wsj.com/articles/bluetooth-ducky-youre-the-one-a-toy-for-more-bath-time-fun-1447341546.
16. FCC (18 December 2014), "17th Report: WT Docket No. 13–135," p. 55. Retrieved from https://www.fcc.gov/document/17th-annual-competition-report.
17. FCC, "17th Report," p. 61. Retrieved from https://www.fcc.gov/document/17th-annual-competition-report.
18. U.S. Bureau of Labor Statistics (28 October 2015), "Price Trends for Wireless and Landline Phone Services, December 2009–September 2015." Retrieved from http://www.bls.gov/opub/ted/2015/price-trends-for-wireless-and-landline-phone-services-december-2009-september-2015.htm.
19. U.S. Congress Office of Technology Assessment (August 1993), *U.S. Telecommunications Services in European Markets* (OTA-TCT-548, NTIS order #PB93-231355), Washington, DC: U.S. Government Printing Office, p. 114. Retrieved from http://ota.fas.org/reports/9349.pdf.
20. International Telecommunications Union (2015), "ICT Facts and Figures." Retrieved from http://www.itu.int/en/ITU-D/Statistics/Pages/stat/default.aspx. There were too many gaps in data to discuss the current state of fixed wire subscriptions. The CEEs are Bulgaria, Croatia, Czech Republic, Estonia, Hungary, Latvia, Lithuania, Poland, Romania, Slovakia, and Slovenia. No data was available with respect to individual mobile phone subscribers for Romania.

21. Calculations are based on the World Bank's country-by-country table of "the sum of active number of analogue fixed telephone lines, voice-over-IP (VoIP) subscriptions, fixed wireless local loop (WLL) subscriptions, ISDN voice-channel equivalents and fixed public payphones" as compiled by the International Telecommunication Union, World Telecommunication/ICT Development Report and database. Retrieved from http://data.worldbank.org/indicator/IT.MLT.MAIN.P2.
22. Tim Greene (19 July 2013), "Black Hat: Top 20 Hack-Attack Tools," *Network World*. Retrieved from http://www.networkworld.com/article/2168329/malware-cybercrime/black-hat--top-20-hack-attack-tools.html.
23. Open Web Application Security Project (OWASP) (17 November 2015), "Mobile Jailbreaking Cheat Sheet." Retrieved from https://www.owasp.org/index.php/Mobile_Jailbreaking_Cheat_Sheet.
24. Allya Sternstein (31 December 2014), "The Cyber Threat in 2015: 10 Twists on Hackers' Old Tricks," *NextGov*. Discussion of trend #5. Retrieved from http://www.nextgov.com/cybersecurity/2014/12/here-are-10-twists-favorite-hacker-tricks-watch-2015/102081/.
25. Allya Sternstein (31 December 2014), loc. cit. Discussion of trend #6.
26. Matthew Hosburgh (2014), "The Spy with a License to Kill." *The SANS Institute*. Retrieved from https://www.sans.org/reading-room/whitepapers/scada/spy-license-kill-35557.
27. Fortinet (2015),"Fortinet Threat Trends in 2015." White paper. Retrieved from https://www.fortinet.com/sites/default/files/whitepapers/Cyber-Security-Threat-Trends-2015.pdf.
28. Fortinet (2015), "Fortinet Threat Trends in 2015." White paper, p. 2. Retrieved from https://www.fortinet.com/sites/default/files/whitepapers/Cyber-Security-Threat-Trends-2015.pdf.
29. PRNewswire (25 March 2015), "No Time to Talk: Americans Sending/Receiving Five Times as Many Texts Compared to Phone Calls Each Day, According to New Report." Retrieved from http://www.prnewswire.com/news-releases/no-time-to-talk-americans-sendingreceiving-five-times-as-many-texts-compared-to-phone-calls-each-day-according-to-new-report-300056023.html.
30. Symantec (April 2015), "ISTR 20: Internet Security Threat Report," p. 25. White paper. Retrieved from https://www4.symantec.com/mktginfo/whitepaper/ISTR/21347932_GA-internet-security-threat-report-volume-20-2015-social_v2.pdf.
31. Symantec ISTR 20 (p. 25).
32. Symantec ISTR 20 (p. 8) cites a Norton survey that reported, "one in four admitted they did not know what they agreed to give access to on their phone when downloading an application. And 68% were willing to trade their privacy for nothing more than a free app."
33. Linda Bustos (2 September 2015), "Cracking Mobile Commerce Myths," Acquia. Retrieved from https://www.acquia.com/blog/cracking-mobile-commerce-myths.
34. J. Gold Associates (February 2015), "Mobile E-Commerce: Friend or Foe? A Cyber Security Study," pp. 3–4. Research paper. Retrieved from https://www.emc.com/collateral/analyst-reports/h13939-mobile-e-commerce-friend-or-foe.pdf.
35. Andy Greenberg (21 July 2015), "Hackers Remotely Kill A Jeep On The Highway—With Me In It," *Wired*. Retrieved from http://www.wired.com/2015/07/hackers-remotely-kill-jeep-highway/.
36. Symantec ISTR 20 (p. 8).
37. Symantec ISTR 20 (p. 10).
38. OWASP (2014), "Top 10 Mobile Risks—Final List 2014." Retrieved from https://www.owasp.org/index.php/OWASP_Mobile_Security_Project#tab=Top_10_Mobile_Risks.

CHAPTER 3

# Blurred Edges: Fixed and Mobile Wireless Access Points

## THE HACKING END GAME

Chapter 2, Wireless Adoption, presented a number of successful hack attack strategies that can help the malicious achieve their goals and that present challenges to defenders. The back stories for these and other evolving attack strategies presented in this and subsequent chapters can be instructive. To deconstruct these and other attacks and how they play out in specific contexts, it is useful to attempt an understanding of the motivation behind them by posing the questions: Why do hackers hack? What are the desired outcomes?

The back stories are as varied as the targets and techniques. Numerous academics, researchers, and industry analysts have attempted to define hacker profiles. According to the UN Hacker Profiling Project (HPP), a joint initiative between the UN Interregional Crime and Justice Research (UNICRI) and the Institute for Security and Open Methodologies (ISECOM), profiles have shifted since the 1970s. Opportunistic lone wolves and script kiddies hacking for knowledge, curiosity, or mischief are being replaced by structured groups and cyber mercenaries with a focused, long-term perspective for achieving desired results. Thus the study sees APTs and more sophisticated activities carried out by malware factories, political hacktivists, and cyber mercenaries working on behalf of nation-states, industrial interests, and organized crime.[1]

Although divisible into multiple subsets, the basic motivations of the threat agents and those who manage them—including those classified as ethical hackers—resonate with those that motivate most human actions: gain, pain, and fear. Table 3.1 shows the relationships between different motivations, end games, manifestations, and hacker profiles. Hackers can, of course, act in response to a combination of motivations. The table indicates some of this overlap. A specific incident may represent a combination of these factors.

Although financial gain frequently comes to mind first, a hacker can also measure "gain" as increases in competitive advantage, intellectual property, reputation, and privileged access (e.g., to online entertainment, system



Hacking Wireless Access Points. DOI: http://dx.doi.org/10.1016/B978-0-12-805315-7.00003-6

**Table 3.1** Hacker Motivation, End Game, Manifestation, and Profile Type

| Motivation | End Game | Manifestation | Profile Types |
|---|---|---|---|
| Gain | Fame/recognition<br>Sabotage<br>Monetary<br>Competitive advantage<br>Espionage | ■ Enhance hacker's reputation<br>■ Ransom access to data or IT systems<br>■ Damage target's reputation/credibility<br>■ Steal identity and other credentials<br>■ Siphon off financial assets (direct/indirect)<br>■ Discredit competition<br>■ Steal trade secrets and other intellectual property<br>■ Traffic in sex, porn, drugs, digital media, weapons<br>■ Use social engineering tactics | Hacktivist<br>Lone wolf<br>Cyberbully<br>Corporate/industrial spy<br>Government agent<br>Dark or Deep Web denizen<br>Cyber criminal<br>Digital media pirate<br>Con artist<br>Script kiddie<br>Cracker |
| Pain | Revenge<br>Sabotage<br>Vulnerability assessment | ■ Transmit malware<br>■ Damage target's reputation/credibility<br>■ Steal intellectual property<br>■ Launch C&C scripts<br>■ Perform covert surveillance<br>■ Perform penetration testing<br>■ Perform reconnaissance activities<br>■ Compromise system resources (e.g., DoS) | Corporate spy<br>Disgruntled employee<br>Cyberbully<br>Cyber stalker<br>Bot herder/Bot master<br>Caller ID/other ID spoofer<br>Ethical hacker<br>Cyber mercenary |
| Fear | Vulnerability assessment<br>Counterhacking | ■ Inject command-and-control malware (e.g., Stuxnet)<br>■ Eavesdrop on government organizations<br>■ Compromise critical infrastructure<br>■ Launch distributed DoS | Ethical hacker<br>Government agent<br>Military hacker<br>Cyber warrior<br>Nonstate actor<br>Paranoid, skilled hacker |

resources that include distributed computing power, ICS). Monetary gain is realized through fraud, accessing financial assets illicitly (often by impersonation), and buying and selling information (e.g., social security numbers, account numbers, medical credentials) to others in the hacking or criminal communities. Ransomware is another modality: threatening to disclose information, destroy, modify, or deny access to information if a cash equivalent payment is not made. Industrial spies realize financial gain by penetrating a company's information resources and obtaining proprietary information and intellectual property. Sale of this information can be arranged on a contract basis or to the highest bidder on a speculative basis.

Pain as a motivation is demonstrated by causing pain, as well as relieving or avoiding pain. Thus, ethical hackers are motivated primarily by relieving or avoiding pain for others through the identification and often mitigation of vulnerabilities. Clearly, they have the skills to pursue financial and other forms of gain. Fear as a motivation is demonstrated by actions like using active cyber defense as a preventive mechanism. For such a mechanism to be effective, at least at the nation-state level, the defender must communicate credibly its limited tolerance for attacks by others, signal willingness to retaliate successfully and appropriately (i.e., proportionately), and be able to deconstruct triggering events to justify responses.[2] The 2007 massive Russian cyberattacks against private and public sector organizations in Estonia inspired fear among Estonian citizens, who are highly dependent on digital communications for banking and news reports.[3] It is easy to imagine how to create citizen fear or distrust of current governmental or industrial reliability for the hacker by other acts of disabling or degrading critical infrastructure sectors used by the public (e.g., water, electricity, transportation).

To achieve their end game, hackers must gain access to information assets. Even when physical access is the primary stratagem used by the attacker, it is often enabled or intensified by first performing reconnaissance to determine what weaknesses exist in the protection of these assets—and how to exploit them. Fixed and mobile WAPs often provide convenient entryways for this discovery.

## DIFFERENTIATION BETWEEN FIXED AND MOBILE WAPs

If we start with the premise that any device capable of receiving and transmitting radio frequencies can be considered a WAP—even if being used as a relay station—the number of fixed WAPs is significant. For our purposes, we will define "fixed" WAPs as those with a long-term commitment to place and geographic location, as opposed to those fixed in situ, but not in loco. A car's factory-installed navigation system, for example, is fixed in situ behind the dashboard; one would not remove it after parking the car for use as an electronic guide while hiking in a national park. And yet, the system is not fixed in loco: Its geographic location changes as the vehicle travels down the road. Meanwhile, a portable navigation system when used in a car is not necessarily fixed in place or geographic location for the long term. Similarly, individuals can create their own mobile wireless hotspot through personal devices, rather than relying on wireless routers that are fixed in place.

## HACKING OPPORTUNITIES IN HYBRID NETWORKS AND COMMUNICATIONS CHANNELS

Wireline connections are less "leaky" than wireless connections due to the innate characteristics of the network media over which signals travel. Still, copper used as the network media is vulnerable to electromagnetic interference (EMI) within the environment from intense, electrical energy sources like motors, transformers, and fluorescent lights. Copper is also subject to crosstalk interference, which occurs when wires are bundled together.[4] A copper-wire connection can be tapped by exposing the internal wires and connecting them to a listening device (a phone or less perceptible "bug"). The tap can be located within the target phone or anywhere along the phone line, including lines on a telephone utility pole. The latter approach is also used to hijack phone connectivity to save the expense (or inconvenience) of subscribing to individual phone service. Of course, such intentional eavesdropping requires physical access to the wire—and is prohibited under Federal and state law (with some specific exceptions).[5]

Businesses and organizations of any size, as well as individuals, may experience wire-based service compromise that generates excessive billing for phone service. Computer users can unintentionally approve a modem hijacking by clicking on web-based ads for "free" content or services.[6] The computer's connection to a legitimate modem is reprogrammed to connect via an international or for-fee phone number.[7] Other phone scams are perpetrated by individuals who call to promote various services (charity donations, credit card assistance, extended car warranties, travel packages) and by robocallers (automated dialing techniques using prerecorded messages rather than live operators).[8]

Fiber optic media, as opposed to copper, are not susceptible to EMI or RF interference and crosstalk. They may, however, be less resilient to physical damage if not adequately protected in conduit, and may be tapped by splicing fiber strands. Coaxial and unshielded twisted pair (UTP) cable are susceptible to EMI interference and can also be physically compromised for eavesdropping.[9] At the physical layer, even transmissions over wired lines can be intercepted, although they are more contained than transmissions over wireless radio frequencies. The important thing to acknowledge is that the vulnerabilities of wired systems carry over into wireless systems. The latter systems also add their own exploit flavors. When considering how to hack (or protect) any communications systems, the characteristics of each element—network media, protocols, architecture, connected devices, applications, physical location, data, users, policies—become part of the calculation for attack (or protection).

## CHALLENGES FOR SECURING HYBRID NETWORKS AND COMMUNICATIONS CHANNELS

Wireless exploits and their tools have generated numerous catchy nicknames and trademarks, often based on the particular 802.11x protocol targeted, which include the following:

- Aircracking—password cracking tool used against Wired equivalent privacy and WPA protections; captures wireless packets to recover password using Fluhrer, Mantin, and Shamir (FMS) attack[10]
- Airjacking—tool for injecting forged packets to support a MITM or DoS attack[10]
- Bluejacking—sends unsolicited, often anonymous, messages over Bluetooth to Bluetooth-enabled devices; messages may contain a vCard (typically for connection to another Bluetooth-enabled device via object exchange (OBEX) protocol; uses include bluedating and bluechatting)[11]
- Bluesnarfing—unauthorized access of information from one Bluetooth-enabled device by another
- Caller ID spoofing—falsifying the caller ID to a number other than the actual calling station's[12,13]
- Commjacking—intercepting transmissions between any device and the Wi-Fi or cellular networks to which it is connected[14]
- Drone skyjacking—drone engineered to take control of other drones within wireless or flying range
- Juice jacking—gaining user access to phone and its contents while it is being charged over a public kiosk using a common USB connection (can compromise privacy and lead to malware injection)[15]
- Skyjacking—exploiting over-the-air provisioning (OTAP) protocols to trap WAPs into connecting to a rogue wireless LAN controller (WLC) or access point; works by transmitting fake radio resource management (RRM) messages with information about the fake WLC; supporting tools include packet injection software[16]

## IMPLICATIONS FOR CONNECTIONS WITH WIRED (LEGACY) NETWORKS AND SYSTEMS

The attacks mentioned above are carried out at different layers of the entire communications system and so are included in the taxonomy of network and device attacks that affect all information systems, regardless of whether they are wired, wireless, or hybrid. Any wired network that can be accessed through a WAP, even if that access is indirect (e.g., an approved desktop that has been used to synch calendars with a compromised wireless device), may be susceptible. Although less colorful than the names applied to the foregoing wireless attacks and tools, the following taxonomy of attacks from Lisa Phifer (published by TechTarget)[17] corresponds to familiar categories and information security principles (e.g., the confidentiality, integrity, and availability—or CIA—triad; the authentication, authorization, and accounting—or AAA—triad):

- Network penetration attacks
  - war driving, rogue access points, ad hoc associations, machine address code spoofing, 802.1x remote authentication dial-in user service (RADIUS) cracking
- Confidentiality attacks
  - eavesdropping, wired equivalent privacy key cracking, evil twin AP, AP phishing, MITM
- Integrity attacks
  - 802.11 frame injection, 802.11 data replay, 802.1x Extensible Authentication Protocol (EAP) replay, 802.1x RADIUS replay
- Availability attacks
  - AP theft, Queensland DoS, 802.11 beacon flood, 802.11 associate/authenticate flood, 802.11 Temporal Key Integrity Protocol Message Integrity Check exploit, 802.11 deauthenticate flood, 802.1x EAP-Start flood, 802.1x EAP-Failure, 802.1x EAP length attacks
- Authentication attacks
  - application login theft, domain login cracking, VPN login cracking, 802.1x identify theft, 802.1x password guessing, 802.1x Lightweight Extensible Authentication Protocol guessing, 802.1x EAP downgrade

## Network Penetration Attacks

As in conventional wired networks, penetrating the network is often number one on the attacker's to-do list. The trick with a wireless network is that defining the network's perimeter (or parameters, for that matter) is complicated by radio frequency signal leaks (i.e., the ability to receive signals from outside a trusted building or other assumed enclosure). War driving—picking up WAP signals using a mobile device—can be a great learning tool for novice wireless attackers (or defenders). Free software tools are available for installation on common platforms. The tools identify and record device characteristics like name, MAC address, SSID, encryption technology, AP manufacturers, geo-location. From this information it is possible to determine where an unprotected or inadequately protected signal is coming from and, if the default device registration name has not been changed, the likely default password for that device.

Rogue access points are unauthorized installations that may be attributed to a trusted user—or an opportunistic attacker. They function in a wireless installation as a wiretap does in one that is wired. Ad hoc associations, on the other hand, represent a different potential for a kind of rogue access point. Peer-to-peer networking, facilitated through SSID broadcasting by wireless devices, is often a default setting that is left enabled. Accidental, incidental, and malicious eavesdropping can result.

## Confidentiality Attacks

Confidentiality is the first element of the security triad and the one associated with privacy: limiting privileged information to those who own it or have an owner-acknowledged need to know it. Concerns about the confidentiality of personally identifiable information have led to enactment of numerous laws globally. Among them are the European Data Protection Directive (1995), the Australian Privacy Act (1998), the US Children's Online Privacy Protection Act (COPPA, 1998), and the Canadian Personal Information Protection and Electronic Documents Act (PIPEDA, 2000).[18]

## Integrity Attacks

Integrity is the second element of the security triad and the one associated with the reliability of information: the assurance that the content and/or associated metadata has not been tampered with. Integrity has not received the kind of attention that incidents involving confidentiality (e.g., data breaches) have received. And yet, compromise of data or message integrity is at the core of some of the most disturbing attacks involving critical infrastructure (e.g., Stuxnet) or critical systems (e.g., Jeep engine hacking demonstration). Data integrity is also critical in medical environments, especially with increased reliance on electronic medical records (EMRs) to minimize undesirable outcomes (e.g., harmful drug interactions). Within the legal sector, the integrity of metadata is critical for date, attribution (or provenance), and time "stamping" of information events and document versions.

## Availability Attacks

Availability is the third element of the security triad and the one associated with the reliability, accessibility, and performance of computing resources (e.g., communication networks, data processing applications, servers) and data. DoS attacks have historically been among the most disruptive for large numbers of individuals and organizations. For critical industrial infrastructure sectors like energy and water, the availability of systems that manage physical controls of distribution networks and pipelines is the most important one of the CIA triad. And in a medical context, this dependency on the digital availability of equipment and information (again, those EMRs) is a life-and-death matter for individuals.

## Authentication Attacks

Authentication is associated with artifacts that validate the identity of the claimant (individual, system, or process) that is attempting to access an information asset. Various credentials may be presented for authentication. Credential theft and compromise are leading factors in data breaches. Concerns about the inherent vulnerabilities in magnetic stripe credit cards, for example, ultimately forced US companies to adopt Europay, MasterCard, and Visa (EMV) technology for the embedded chip credit cards rolled out in 2015. The theft of credentials is a key factor in multilayered attacks—by allowing privileged access to key host systems—as well as in credit card fraud, identity theft, and unauthorized building access.

## C&C Attacks on Automated Processes

Most networks and attached devices are based on a set of automated (i.e., C&C) processes. The ultimate prize for a hacker is to obtain control over the C&C processes using a combination of the above attacks, which then allows the hacker access to all resources, meanwhile either masking or hiding evidence of that activity. C&C servers can continue working for years and are no longer limited to botnet activity or Internet relay chat servers to direct victims. They have been found among cloud infrastructure service providers and a range of Internet domains. Table 3.2 indicates known malware families that reside in C&C servers that affect cloud infrastructure services.

C&C attacks against critical infrastructure elements also take advantages of known vulnerabilities in legacy software that cannot be patched or updated reliably.

**Table 3.2** Malware Families with Command and Control (C&C) Servers on Cloud Infrastructure Services[a]

| Malware Family | Description |
|---|---|
| ZeuS | Cybercrime |
| POISON | RAT, targeted attacks |
| CLACK | Adware |
| BOZOK | RAT, targeted attacks |
| IXESHE | Targeted attacks |
| ESILE | Targeted attacks |
| DUNIHI | RAT, targeted attacks |
| KELIHOS | Cybercrime |

[a]*Marco Dela Vega (14 September 2015), "How Command and Control Servers Remain Resilient,"* TrendMicro: Trend Labs Security Intelligence Blog. *Retrieved from https://blog.trendmicro.com/trendlabs-security-intelligence/adapting-to-change-how-command-and-control-servers-remain-hidden-and-resilient/.*

## RECOMMENDATIONS FOR WIRELESS/HYBRID SYSTEMS

The challenges of securing wireless technologies should not obscure their advantages. Fixed wireless access connections are now widely regarded as desirable in terms of installation cost and ease,[19] especially in conditions where access to the cable or wiring plant or plenum access is difficult or restricted, time schedules are tight, few walls or other environmental obstructions exist, no competing telecommunications infrastructure is in place, or coverage areas are widely separated or underutilized. The cost of protecting wireless communications, however, must be calculated when building the business case for a wireless or hybrid system. The challenges simply speak to the importance of understanding what your assets are, where they are located, who has access to them, how they are being accessed, and how they should be protected. The objective of such situational awareness is to be aware of what should be and what is. The objective is also to be aware of what could be, thus learn to think like an adversary as Sun Tzu advised: "To know your Enemy, you must become your Enemy."

### A VERY BRIEF HISTORY OF EARLY CIVILIAN WIRELESS USE

During the pre-World Wide Web era of the 1960s, the Arpanet backbone connected universities over wired, POTS networks: conventional, point-to-point networks. By the 1970s, the concept of long-haul data connectivity had been proved and adopted to the extent that researchers and analysts recognized the cost and performance inefficiencies of these dedicated, unshared connections. Spread spectrum capabilities like frequency hopping that would allow shared connection were appealing in the abstract for civilian uses and identified as a possibility by researchers at the University of Hawaii in 1968, but the technology needed more development. A key driver behind the effort launched there for "an experimental radio linked computer network" was to link satellite campuses (an undergraduate facility in Hilo and five 2-year community colleges on three other islands including Hawaii) with the Honolulu-based main campus's computing resources.[20]

The project showcased the use of radio for computer internetworking and work continued to perfect the model as supporting software protocols were developed and regulatory constraints on frequency allocations were addressed. As packet-switching technology evolved, it became a viable competitor to circuit-switching and was recognized for enabling certain economies of scale.[21]

### Know Your Enemy

By looking at your environment from the perspective of a likely attacker, priorities about where to invest in protection efforts become clearer. Considerable investment may be necessary to achieve a successful attack. The rational attacker will make a deliberate calculation about what constitutes success and the desired return on investment (in terms of time, tool, financial, and skill resources) and weigh these against the probability of success—and the probability of failure or adverse outcomes (e.g., discovery, legal conviction, professional ostracism). Understanding the end game of potential adversaries is the first step to modeling which attack vector(s) are likely to be the most profitable or least costly.

This analysis assumes intentionality and sophistication rather than opportunistic digital vandalism. The latter category can include acts as insipid as keylogging and remote terminal experiments performed by grad students against staff in a university setting to the devastating Morris Worm, launched in 1988 from MIT by Cornell grad student Robert Morris. Estimates vary as to how many computer systems were crashed as a consequence, but 10% of systems connected to the Internet is often cited in articles. At that point in time, about 50,000–60,000 computer systems were connected to the Internet.[22] (As a point of reference, a 2015 report estimated that 1.2 billion devices were connected to the Internet; 10% of that would be 120,000,000 devices.) Morris earned fame for his eponymous malware, in addition to 3 years' probation, a fine for his intentional unauthorized access of others' computers, and conviction under the federal Computer Fraud and Abuse Act[23] (enacted in 1986 as an amendment to the 1984 Counterfeit Access Device and Abuse Act).[24]

Acts of digital vandalism perpetrated over wireless systems include object lessons designed to create awareness about unsafe practices. A regular feature at the hacker conference DefCon is the large video display, the Wall of Sheep. These are the accounts compromised through the Wi-Fi hotspot by transmitting usernames and passwords in clear text. (It is an educational conference, after all.) This public shaming is, in effect, close captioned, and has not carried the same fatal results as have public shaming through social media bullying.

As a defender of your network and assets, you should:

- Monitor traffic details and understand behavior patterns to determine what is normal
- Recognize anomalous behavior on the network and investigate
- Quickly assess and share relevant traffic and contextual information with security resources
- Ensure your network has active tools for consistent sensing and detection of both intense and low-level (but persistent) attack or reconnaissance activities, in addition to robust information processing, anomaly detection, signaling, reporting, and dissemination
- Perform ad hoc and scheduled "attacks" on your network to determine the effectiveness of your defenses and design improvements to those defenses

## IMPORTANCE OF CONTEXT (USE CASE SCENARIOS)

Initially considered an expensive alternative to conventional circuits, and thus relegated to implementation as a backup to wired communications, wireless is an important feature of most communications infrastructure being built today. As discussed in Chapter 2, Wireless Adoption, wireless connectivity has become the preferred choice in countries that have not historically deployed significant wired infrastructure. Without the legacy of managing communications security and following safe information handling practices, the protection of information assets in these countries may become more complicated.

Although the academic exercise of planning an attack or its defense can be entertaining, looking at real-world instances gets to the operational, legal, regulatory, and behavioral environments in which WAPs are found. The chapters that follow explore the characteristics of specific applications for WAPs in a range of contexts: individual (personal, social, recreational), commercial (retail, multifamily, office).

These use case scenarios will attempt to capture insights about where the value resides that meets an attacker's preference for acceptable risk, exploitable attack surfaces, planning/executing time horizon, available tools/skills, reasonable cost, and sufficient payoff. By analyzing attacker behavior in terms of market opportunity and the rational drive to maximize return of investment/effort, we can arrive at a better understanding of how to adapt prevention, detection, and control mechanisms to specific real-world contexts.

## ENDNOTES


1. Raoul Chiesa (23 April 2015), "To Be, Think and Live as a Hacker: The Hacker's Profiling Project of the United Nations (UNICRI)," Presentation at DKR, Danish Crime Prevention Day, April 23, 2015, Aalborg, Denmark. Retrieved from http://www.dkr.dk/sites/default/files/raoul_chiesa_dkd15.pdf.
2. Emilio Iasiello (Autumn 2014), "Hacking Back: Not The Right Solution," *Cyber Strategies*, 105–113, Strategic Studies Institute. Retrieved from http://www.strategicstudiesinstitute.army.mil/pubs/Parameters/Issues/Autumn_2014/13_IasielloEmilio_Hacking%20Back%20Not%20the%20Right%20Solution.pdf.
3. Scheherazade Rehman (14 January 2013), "Estonia's Lesson in Cyberwarfare," *U.S. News and World Report*. Retrieved from http://www.usnews.com/opinion/blogs/world-report/2013/01/14/estonia-shows-how-to-build-a-defense-against-cyberwarfare.
4. Greg Tomsho (2016), *Guide to Networking Essentials, Sixth Edition*, Cengage Learning.
5. Gina Stevens and Charles Doyle (13 January 2003), "Privacy: An Overview of Federal Statutes Governing Wiretapping and Electronic Eavesdropping," *Congressional Research Service*. Retrieved from https://epic.org/privacy/wiretap/98-326.pdf.
6. Verizon (n. d.), "Modem Hijacking." Retrieved from https://www.verizon.com/support/residential/phone/homephone/billing/privacy+and+security/modem+security/96042.htm.
7. Crimes of Persuasion (n. d.), "Cramming of Unauthorized Service Charges, Membership Fees, Subscriptions or Payments on Bank Accounts, Credit Cards or Telephone Bills Fraud." Retrieved from http://www.crimes-of-persuasion.com/Crimes/Telemarketing/Inbound/MinorIn/cramming.htm.

8. US Federal Trade Commission (n. d.), "Phone Scams." Retrieved from http://www.consumer.ftc.gov/articles/0076-phone-scams.
9. Kenneth Mansfield, Jr., and James L. Antonakos (2009), *Computer Networking from LANs to WANs: Hardware, Software, and Security.* Cengage Learning.
10. Infosec Institute (23 February 2015), "13 Popular Wireless Hacking Tools." Retrieved from http://resources.infosecinstitute.com/13-popular-wireless-hacking-tools/.
11. Multiple discussions, for example, those retrieved from Wikipedia https://en.wikipedia.org/wiki/Bluejacking and Professor Messer's Comptia Security+ textbook http://www.professormesser.com/security-plus/sy0-401/bluejacking-and-bluesnarfing-2/.
12. Andrew Swoboda (20 April 2015), "How to Protect Yourself From Caller ID Spoofing," *Tripwire.* Retrieved from http://www.tripwire.com/state-of-security/security-awareness/how-to-protect-yourself-from-caller-id-spoofing/ and Robert Lemos (4 August 2015), "Trust No One: How Caller ID Spoofing Has Ruined the Simple Phone Call," *PC World.* Retrieved from http://www.pcworld.com/article/2951748/security/trust-no-one-how-caller-id-spoofing-has-ruined-the-simple-phone-call.html.
13. Alas, free tools like SpoofCard <https://www.spoofcard.com/> are advertised on the Internet.
14. Ben Kepes (12 May 2015), "CoroNet Launches to Put a Stop to 'Commjacking,'" *Forbes.* Retrieved from http://www.forbes.com/sites/benkepes/2015/05/12/coronet-launches-to-put-a-stop-to-commjacking/ and Lee Badman (18 May 2015), "Security Startup Defends Against WiFi Hijacking," *Information Week.* Retrieved from http://www.networkcomputing.com/wireless-infrastructure/security-startup-defends-against-wifi-hijacking/a/d-id/1320438.
15. Tech Advisory.org (4 September 2014), "What's juice jacking?" Retrieved from http://www.techadvisory.org/2014/09/whats-juice-jacking/.
16. AirTight Networks (n. d.), Webinar. Retrieved from http://www.airtightnetworks.com/fileadmin/pdf/webinar/Skyjacking_FAQs.pdf.
17. Lisa Phifer (June 2009), "A list of wireless network attacks," *TechTarget.* Retrieved 12/22/15 from http://searchsecurity.techtarget.com/feature/A-list-of-wireless-network-attacks (Note: Includes excellent tables with attack descriptions and methods/tools as well as a wireless network vulnerability checklist, retrieved from http://searchsecurity.techtarget.com/feature/A-wireless-network-vulnerability-assessment-checklist).
18. A useful reference for laws governing data protection in 25 different national jurisdictions around the globe is *Data Privacy and Protection* (September 2015), edited by Rosemary P. Jay and published by Getting the Deal Through. Retrieved from https://gettingthedealthrough.com/area/52/data-protection-privacy/.
19. Tom Harris (n. d.), "How Wiretapping Works." Retrieved from http://people.howstuffworks.com/wiretapping3.htm.
20. Norman Abramson (March 1985), "Development of the ALOHANET," *IEEE Transactions on Information Theory*, Vol. I-31, No. 2, 119. Retrieved from http://www.eletrica.ufpr.br/mehl/te155/abramson-ALOHANET.pdfcomputing.
21. Ibid., p. 122.
22. Interview with Eugene Spafford (25 October 2013),"Lessons From the First Major Computer Virus," *Intel Free Press.* Retrieved from http://www.intelfreepress.com/news/lessons-from-the-first-computer-virus-the-morris-worm/7223/.
23. Larry Seltzer (2 November 2013), "The Morris Worm: Internet Malware Turns 25," *ZDNet.* Retrieved from http://www.zdnet.com/article/the-morris-worm-internet-malware-turns-25/.
24. Although computer viruses had existed at least five years prior to 1988, the Morris Worm's release served as a wake-up call to governments, technologists, and individuals. DARPA created the US computer emergency response team coordination center (CERT/CC) at Carnegie Mellon University to address similar challenges in the future.

CHAPTER 4

# Hacks Against Individuals

How attractive a target are you to a potential hacker? No offense, but as an individual, it is unlikely that you are very attractive to anyone but a low-level cyber vandal, pickpocket, or identity thief—unless you are a celebrity, politician, high-profile executive or government official, security guru, or high net worth individual (HNWI). Still, the damage and inconvenience that can be triggered by these opportunists can be considerable. In many cases covered by security industry specialists and media, social engineering was integrated into the attack process. In a complex attack that combines technical expertise with social engineering, the latter is most critical in the reconnaissance phase. It is here that social engineering can save time, effort, and potential harm to the one attacking. Talking one's way into a privileged area (whether physical or digital), for example, is typically less dangerous than physically breaking into an office or home. Breaking and entering at the physical level is detectable and carries legal consequences—and risk of actual physical injury—as opposed to gently conning people. Combining social engineering with hacking WAPs is highly productive, substantiating the observation that "technical engineering and social engineering go hand and hand."[1]

Perhaps the most significant advantage to social engineering is its low cost and ease of use. It does not require deep understanding of technological tools or programming techniques, just awareness about factors that motivate human behavior and the desire to manipulate others into acting in a way that is beneficial to the social engineer. In this we are all trained as social engineers so, ostensibly, we should be able to recognize when we are being escorted down a path that might be risky. Ostensibly.

A good definition of social engineering comes from McAfee: The deliberate application of deceitful techniques designed to manipulate someone into divulging information or performing actions that may result in the release of that information.[2]

Psychologist Robert Cialdini categorizes the influencing levers for social engineering, that is, the human susceptibility to different types of social



Hacking Wireless Access Points. DOI: http://dx.doi.org/10.1016/B978-0-12-805315-7.00004-8

engineering appeals as reciprocation, scarcity, consistency, liking, authority, and social validation.[3]

*Reciprocation* addresses the human compulsion to return a favor. You offer to hold open the office door for me because I'm carrying a heavy load, and I am inclined to ignore that you have entered the building without having to swipe your access badge.

*Scarcity* captures the tendency to fill a void, often of information (e.g., bogus email messages about missing account or credit payment information). An example of how information scarcity can play out is the 2011 hack of RSA, manufacturer of SecureID tokens, used by some of the most security-conscious organizations in the world and broadly distributed: some 40 million RSA tokens were in use in 2009. Similar software runs on some 250 million smartphones, as of 2011. The hack ultimately affected many organizations and cost RSA $66 million,[4] in addition to significant unpleasant publicity and questions about the cryptographic approaches used to implement multifactor authentication. The phishing email sent to a small group of employees contained an Excel spreadsheet with an appealing subject line: "2011 Recruitment Plans." Although RSA's email filter routed this message correctly to the spam folder in individual mail accounts, a curious employee retrieved and opened it, thus releasing its malicious payload: a zero-day Adobe Flash exploit that then allowed attackers to deploy a version of Poison Ivy RAT, a remote administration tool. This allowed the attackers to control computers from outside the network and obtain access to personally identifiable information (PII) and account credentials.[5]

*Consistency* refers to that human desire to make good on a promise or commitment, the intention to be trustworthy, based on the assumption that others are similarly trustworthy. My sense is that this underlies the often-noted phenomenon of individuals clicking on a link to change a password without verifying the message's source even though that individual has just attended a security awareness training session. Deep learning that training warnings pertained to all messages with links, even those that appear to originate legitimately, has not yet occurred. This technique is useful for evaluating whether training has achieved its goals and where gaps still might exist.

*Liking* is similar to pleasing. People are inclined to trust those they like, admire, or sympathize with—and overshare with them; examples include Bernie Madoff, Frank Abagnale (of "Catch Me If You Can" fame and for 40-plus years working for good with the FBI and others), and James Hogue (the "Princeton Imposter" who is still working cons, recently in Colorado). This human frailty is frequently exploited in voice messages that urge the recipient to some course of action like dialing a masked for-fee phone number.

*Authority* is often used in phishing attacks to gain access or credentials. One example is the UPS fraud that leads the target to a legitimate page for checking shipping status, but then to a faked invoice link that triggers the downloading of malicious payload. Natural credulity contributes to the strength of this particular lever, especially when the caller (or emailer) claims to be from an organization or service that the target frequents. Challenging authority can be a recommended course of action—or at least verifying the identity of the message sender or caller by calling back and/or using one of the reverse 411 or email identification tools available online.

*Social Validation* leverages the crowd-following tendency to participate in the news being shared, ostensibly by a friend. It also takes advantage of reputation-based networks (which are, nonetheless, susceptible to hacking).

Although Cialdini does not discuss greed as a basic social engineering lever, the wish for gain is certainly a factor that is used in many attacks. Many people have apparently not given up on the childhood belief in a fairy godmother, magic bean, or unknown and yet wealthy benefactor. Social engineering techniques are a powerful adjunct to technical tools and can save money, time, effort, and, possibly, detection leading to criminal prosecution. The techniques can be used to acquire the information or credentials needed to support attacks on WAPs. They can also alleviate the need for using more sophisticated tools like brute force attacks against passwords.

WAPs—for example, smartphones, tablets, baby cams, fitness trackers, routers, navigation systems, clothing, household appliances—are like pores in the ubiquitously networked world we are building. And like pores, they allow signals and data in and out, some desirable, others not.

Each hacker scenario that follows indicates the apparent or asserted attack objective on the individual, the technique employed, and the impact to the victim. Included in Appendix 1 are three sample attack diagrams that indicate some possible routes for achieving a particular objective. The diagrams show a possible public WiFi compromise, spear phishing paths, and network access through medical devices.

## ACCOUNT CREDENTIALS CHAIN ATTACK

*Hacker Objective*: Power play; wreak havoc by taking over a security industry analyst's digital life

*Hacker Technique*: Chain reaction after single point of entry via social engineering

*Victim Impact*: Loss of data, especially priceless family photos; faked Twitter posts read by his approximately 131,000 followers

*Wired* magazine reporter Matt Honan experienced an epic hack that exposed the security flaws in Google, Amazon, and Apple ID (iCloud) policies and practices that support user convenience, but also remove digital stopgaps between applications and personally identifiable account information.[6] Although he was ultimately able to recover 100% of his daughter's precious first-year-of-life baby pictures and did not suffer credit card and other financial account compromise, he did lose 25% of his files, numerous applications and account preference settings, and a week of productivity while working anxiously to learn what had happened and what, if anything, could be restored from the erased memory of his iPhone and MacBook Air. He also had to repair his Twitter account, now bowdlerized with racial slurs and other offensive commentary, and pay more than $1600 to a data recovery firm (an investment he was happy to make).[7]

## Victim's Chronology

1. iPhone powered down by itself. When plugged, displays setup screen.
2. iCloud password will not work. When iPhone plugged into computer, user informed that Gmail account information was wrong.
3. Called AppleCare. Tech did not mention an earlier call from an impersonator who claimed to have had problems getting into his Me.com email account (which the impersonator had guessed, based on Gmail account listed on personal web page). Impersonator was given a temporary password in spite of not knowing security answers and just providing two pieces of information readily available: billing address and last four digits of the registered credit card.
   a. Billing address available by checking "whois" on personal web domain. Alternatively, sites like Spokeo, WhitePages, and PeopleSmart are resources.
   b. Credit card discovery required extra steps.

As a self-respecting hacker with limited patience for the messiness of breaking into an office, burglarizing a home, or breaking an automobile window to obtain mobile computing devices (high-probability repositories for contact lists, user account information, and access credentials), an attacker who was involved in Honan's account compromise (Phobia) contacted him to, in part, bruit that he had used social engineering, nonphysical techniques to break into the account. He just called tech support at AppleCare and provided Honan's billing address (easily deduced from an online search for publicly available information about Honan) and the last four digits of a credit card. No brute force efforts for password cracking or rainbow tables for reverse engineering cryptographic hashes were needed, nor was dumpster diving to obtain discarded bills revealing the last four digits of credit cards.

## The Final Four: Credit Card Attack Surface

This hacker took advantage of retailer eagerness to provide a prospective customer with convenient service. By claiming to be an Amazon account holder to the Amazon representative he called, then providing easily accessible information (name, email address, billing address), he was able to "add" a credit card number to the account. For ease of use and less traceability to yourself, you can obtain "valid" credit card numbers (plus card verification value (CVV) codes) for free from a number of websites, although you might have to confirm that you will not use the number obtained for illegal purposes. (*Hacker Response*: "OK.") You can also use an expired or retired credit card (your own or someone else's) and just change the expiration date.[8]

Having thus established credibility with tech support, the hacker then called back to say he had lost access to the account. By using the new credit card as validation (as well as the name on the account and billing address), you can add a new email account, then request a password reset to the new account. You will then have access to the account and be able to see, in clear text, the last four digits of all credit cards associated with the account.[9] Using this additional information you can work with trusting tech support for AppleId to respond correctly to identity verification questions based on the last four digit of the credit card used to subscribe to the service. Once you-the-hacker have succeeded, you can remotely wipe files from the victim's iPhone and other connected Apple wireless devices, change access codes, and even alter file information or obtain credentials for purchased apps and the Apple Store.

Similar processes can be used for accessing Google accounts. Because Google integrates account sign-ons across product platforms, by compromising authentication credentials in one product all are compromised. Thus, Honan had to reconstruct his identity with Google support. This time he opted for multifactor authentication to stymie would-be hackers and also revoked connection permissions to his Google account for all apps and websites.[10] He also no longer uses his Apple email as the backup contact account.

As an aside, the EMV cards, required since 1 October 2015 for compliance with revised standards from major US credit card issuers,[11] are more resilient to attack for in-person transactions. In card-not-present situations, such as those associated with most wireless e-commerce activity, the embedded chip protection is irrelevant.

## Takeaways

This attack shows that even when one "does everything right," a person may still be hacked. Information on the phone would have been lost had the phone been stolen as opposed to having been wiped. Backing up information on a phone to another device (or to the cloud) is recommended, but few people take the time to perform this—or only start doing it after the loss of photos or other items.

## PUBLIC WI-FI HOTSPOT ATTACK

*Hacker Objective:* Obtain account credentials and device access

*Hacker Technique:* Deploy a rogue WAP and hijack communication signal or compromise the legitimate wireless router

*Victim Impact:* Potential loss of data, credential misuse, botnet recruitment, ransomware hard drive access block

### Victim's Chronology (Composite Description)

1. Opens wireless utility and scans available networks. Selects the logical network name.
2. Performs usual activities, for example, online banking, medical record check, e-commerce purchases, remote alarm or thermostat control for home.
3. Notices computing performance degradation, multiple spam or suspicious email messages, alerts from friends about odd messages received, questionable credit charges or banking account activity, lockouts of entertainment-related applications (e.g., iTunes, Pandora, Netflix).
4. Sees evidence of tampering with baby monitors, thermostats, security alarm systems, etc.

### Hacking Technique: Honeypot Look-Alike

Computing devices, like their human users, are programmed to work efficiently and take the path of least resistance (or, at least, the path that leads to less battery usage and/or faster promised connection speed). A low-key, opportunistic attacker can set up a rogue WAP with an innocent or expected name, for example, Pablo's Danger Monkey.[12] If the rogue WAP is in closer proximity or broadcasts a stronger signal to devices that are looking for a wireless connection, the device will likely choose that open, unprotected connection, assuming the user does not take the time to launch a VPN session. As one analyst describes the encrypted tunnel established between computing device and ISP during a VPN session:

> Think about it this way: if your car pulls out of your driveway, someone can follow you and see where you are going, how long you are at your destination, and when you are coming back. With a VPN service, you are essentially driving into a closed parking garage, switching to a different car, and driving out, and no one who was originally following you knows where you went.[13]

Keep in mind, however, that your device is vulnerable and signals are unprotected until after the VPN is established.

## Hacking Technique: Commjacking (Exploiting Public Wireless Router Attack Surface)

Coffee shops are prized for the quality of their brew, not the security of their networks; technology investments are centered on finding the best value coffee-brewing equipment. Baristas are hired for customer service skills and earn money for the shop by being out front, not in the back with whatever computing technology is available. Logically, there should be no expectation on the part of customers that the technology is adequately safe for conducting any activities that include communication of PII, financial accounts and credentials, or intellectual property. Buying and deploying a wireless router is a low-cost approach to providing customer (dis)service unless the router is hardened: default factory settings are changed, administrator accounts are differentiated and monitored by user, encryption is ensured to at least WPA2, and passwords and network performance and activity metrics are monitored regularly.

A rogue WAP can be deployed that rides on the communication signals of the legitimate network and establishes a credible presence to which customers willingly connect. Another ploy is to take control of the wireless router if it has not been sufficiently hardened. In the latter situation, the connecting mobile device may, understandably, recognize that the router is familiar and trusted. "MITM" attacks are thus easily carried out, using low-cost tools like Hak5 WiFi Pineapple (less than $100 for a pocket-size wireless penetration testing/auditing tool).

## Takeaways

VPNs are typically provided by businesses to their employees. Even with VPNs, employees are discouraged from performing business on public Wi-Fi hotspots, since other techniques such as "shoulder surfing" can lead to the loss of information.

One should not perform critical applications, such as banking, on unsecured WAPs. Even if a bank uses two-factor authentication, such as a PIN sent by text to a cell phone as well as an answer to a security question on your tablet, both pieces of information can be sent over the same WAP. (Many cell phones use Wi-Fi hotspots when available to conserve data usage.)

If a public WAP requires users to log in to use the service, a hacker could set up his rogue WAP to provide a victim with a similar page. By logging onto this page, malware could be downloaded which could lead to the victim's device becoming part of a botnet or being infected with ransomware.

Most operating systems will notify a user if a new Wi-Fi network is being accessed and request that the user identify how the network should be considered and used. One should always check with the "owner" of the public WAP on the proper network to be used the first time the network is accessed. A different Wi-Fi network on subsequent uses should be treated with caution and verified with the owner as to its "bona fides."

## THE GULLIBLES TRAVEL ATTACK

*Hacker Objective:* Obtain account credentials, corporate intellectual property, money from (often) HNWI or persons of interest

*Hacker Technique:* APT to compromise a high-end hotel's wireless network and reservation system to deliver malware via P2P and other connections

*Victim Impact:* Loss of data, credential misuse, botnet recruitment, ransom exposure, financial and intellectual property loss

As many as 16 million people fall victim to identity theft in the United States annually, with international travelers as much as three times more likely to experience such problems. Even sophisticated travelers can be caught unawares.

"Darkhotel" is a variety of malware identified in at least 3000 targeted attacks on high-level corporate business executives staying in hotels while traveling. The majority of the attacks (90%) played out in Japan, Taiwan, China, Russia, and South Korea. Investigative security firm, Kaspersky Labs, found that this particular APT was more sophisticated than many, and combined watering hole and spearfishing techniques.

### Victim's Chronology (Watering Hole Technique)

1. Logs onto hotel network without using VPN.
2. Receives email message(s) to upgrade or patch common software applications like Adobe or Google.
3. Applications contain digitally signed backdoors (that are installed along with the legitimate code).

These attacks are multivector. For example, spearfishing techniques are apparent, that is, the targeted zero-day attacks have appeared to specify just certain individuals, as if the attackers knew in advance the scheduled arrival and departure of hotel guests. The hotels affected required both last name and room number to authenticate the guest who was logging on. The suspicion among investigators is, therefore, that both the hotels' wireless networks were compromised along with their reservation system. The attackers also appear to have delegitimized certificates with 512-bit encryption keys.[14]

The investment required for factoring 512-bit encryption keys has decreased dramatically since these keys were first used in 1999: Researchers from the University of Pennsylvania showed that an investment of $75 and four hours would be needed, using the virtualized computing resources available by the hour from Amazon Elastic Compute Cloud (Amazon EC2).[15]

Key length makes a notable difference in the time needed to crack them. Experienced attackers and ethical hackers can also obtain secure shell (SSH) public keys that have been uploaded to GitHub's public API.[16] Legacy technology infrastructure is thus a high-opportunity attack surface: weak encryption, default or unsophisticated privileged account naming/credential protection, unpatched vulnerabilities, and expired vendor or manufacturer support.

As with other near field communication (NFC) devices like chip-enabled credit cards, passports are a potential source for compromise. Almost all US passports issued since October 2006 are embedded with passive RFID chips (the US Department of State's initial plan was for active RFID chips) that contain useful PII: name, nationality, sex, date of birth, place of birth, digitized photo of individual, and soon, digitized biometric data. Other nations have followed suit, in keeping with the UN's International Civil Aviation Organization standards. Hackers can use NFC readers to obtain information on the RFID chips; although some protection is offered by the passport cover, RFID-blocking sleeves, wallets, and other containers can reinforce that protection.[17] (An added advantage to such protective mechanisms is that you could protect your hotel room magstripe card from accidental degaussing even if sharing a jacket pocket or purse with your smartphone.) Of course, the do-it-yourself approach is always cost-effective: wrapping your passport and credit card in aluminum foil means "game over" for malicious card readers.

## Takeaways

Many companies will provide loaner tablets or PCs for employees traveling overseas. These devices will have the latest security updates and connect with corporate offices over a secure VPN (1024-bit encryption or better), and will ensure all functions are "performed in the cloud," with no information allowed to be stored on the device. After returning to the United States, these devices are checked to see what malware has been loaded onto the device to understand what additional security functions are needed. The device is then wiped and re-imaged for future remote travel use.

The US Government's Secure Mobile Computing initiative is based on the premise that a secure mobile device cannot have both secure and nonsecure connections. The secure devices are provisioned so that only one connection is made to an agency's portal and that that connection is secured in a VPN tunnel. The agency portal is responsible for managing connectivity to resources on the Internet.

It is only a matter of time before one or both of the above approaches become "best practices" for industry.

## THE INTERNET OF HACKED THINGS ATTACK[18]

*Hacker Objective:* Credential theft for financial gain; signal access for surveillance, botnet activities, entertainment, man-in-the-browser access

*Hacker Technique:* Use under-protected, networked, wireless devices as a pivot point (entry) to information repositories, other devices, and communications routers

*Victim Impact:* Botnet recruitment, device bricking, loss of data, credential misuse, ransomware exposure, burglary, stalking

### Gadget Attack

Each interconnected consumer gadget with inadequate protection creates a new surface for attack in a kind of smart-device-meets-dumb-user dance: picture an evil and competent Inspector Gadget cyborg[19] that can extend his reach into your pocket, your kitchen appliance, your baby's crib. Although cyber-pranks like hacking into a refrigerator to play with the temperature setting or food replenishment schedule are trivial, using that entry point to gain access to other devices connected through the same home-based infrastructure is not. Because the information gathered by such devices seems incidental, it is easy to dismiss the potential for combining that information, perhaps with publicly shared information communicated over social media networks, to form the basis for carefully designed attacks deployed digitally or physically.

Gartner research analyst Earl Perkins explains it well: "If I look at my home as a bubble, the threat opportunity increases with every hole I put in the bubble and if I'm wearing wearable technology or have video surveillance system linked via the Internet for example, I have all these points of access and that is power for people wanting to steal personal information."[20] One large-scale attack during the 2013–2014 winter holidays (when attackers are especially busy) engaged more than 100,000 common gadgets (especially wireless home routers) in a malicious email campaign that transmitted 750,000 messages. More than privacy violations are at stake, especially given the number of gadgets being connected every day: Gartner projects that the number of connected devices will reach 20 billion by 2020. The software controlling device operations is frequently not accessible by consumers, facilitating the attacker's quest for access to devices inadequately protected by manufacturers.

Wireless consumer-grade surveillance devices like baby cams offer remote Peeping Toms insight into personal lifestyle patterns and, introduce a distinctive creepy factor as familiar home and other environments are remotely and covertly switched into Panopticon-style environments.[21] Researchers who tested Internet-connected baby monitors gave them failing grades on basic security dimensions: lack of encryption, default passwords, accessible identification credentials (device serial and account numbers). Countermeasures include decoupling the monitors from the Internet and just using radio-frequency signaling, which limits interception to those in the immediate vicinity of the device, thus raising the risk to the hacker.[22]

Current generation fitness trackers generally do not take advantage of Bluetooth LE security techniques that mitigate against MITM attacks. Rather, they happily pair with other devices, sharing login credentials and activity tracking information with little data integrity validation for the latter (i.e., the data is not tamperproof and often transmitted as cleartext). Although not considered, medical devices in the United States, thus not covered under HIPAA provisions, such data is considered protected under the European Data Protection Directive. Legal issues with respect to the data generated as evidence in insurance litigation cases is discussed in Chapter 6, WAPs in Medical Environments.[23]

Other low-level nuisance attacks against individuals that may or may not carry a significant monetary damage combine social engineering and other techniques to gain access to WAPs. A few common ones are described below.

## Bricking a Device

Consider the uses to which you could put your smartphone, tablet, or other mobile device if rendered useless as a communication or information device: paperweight, shim for an uneven table, plant stand, hotplate. Basically, if the device is *hard-bricked,* this is the brainstorming list you need because your device is dead: it will not run and is not recognized by your computer, perhaps because of a failed user effort to root the device or a successful external hacking effort against an already rooted device. A device that is *soft-bricked,* on the other hand is not working as it had been (e.g., bootlooping without completing initiation, crashing frequently, or failing apps).

## Your Ship(ment) Has Arrived

Whether playing on individual greed ("At last! A cherished vendor who accidentally sent me a gift!") or forgetfulness ("Was I sleep-shopping online again?"), hackers realize good success with bogus email and even snail messages (often postcards) that encourage the recipient to open up an attachment, provide credit card information, or dial a toll-free number to claim the mystery item. The messages that are secretly inviting you to download malicious code, give up credential information, or slam your telephone number likely appear from a legitimate service provider: DHL, FedEx, UPS.[24] Although the package delivery variants appear more frequently during holiday gift-giving season, rest assured that hacking season is year-round.[25]

## Unique Versus Predictable

Unlike the odds of predicting successfully whether or not you will win a Powerball lottery, the odds of predicting successfully your credit card's security code are high. Magnetic strips (magstripes) containing "unpredictable" numbers ranging from 0 to 99 are still being used on card readers (again, convenience trumps security: legacy systems will remain intact, although their vulnerability is acknowledged). Given a single attempt, the likelihood of guessing the number would be one in 100—hardly a random number, especially given the speed at which simple computers can perform matching exercises. In addition, databases of actual credit card numbers (including scores that indicate the associated credit limit) are available online,[26] as are credit card number generating tools.[27] The latter tools should also generate a CVV and expiry date. Continuing to rely on legacy magnetic stripes perpetuates known vulnerabilities that have been famously exploited in the past.

Magnetic stripe technology is limited with respect to implementing "fixes like including more complex random numbers into transactions to prevent card cloning."[28] In fact, an Oklahoma State University student cloned a fake credit card for use on campus as a final project before graduation by deconstructing the pattern in the "random" 16-digit numbers assigned by the school. Of course, his same technique could be used to create fake building access credentials, thus potentially introducing physical as well as financial safety issues.[29] Differentiating between "unique" and "predictable" has also entered into discussions about whether government agencies and security technology firms have reached agreements on sharing algorithms for generating what are actually pseudo-random numbers. Computers follow programmed guidelines; true randomness does not.[30]

## Digital Pickpocketing and Data Slurping

Credit card chips are passive and require activation by smart card readers. To be able to skim information from these cards, specialized technology (card readers) must be purchased, so some investment is required by the potential hacker. The investment is not onerous, however: less than $100 through online vendors. Contactless cards are a different challenge because they rely on RFID technology. European standards recommend that the RFID chips be readable at a maximum distance of 5 cm readability, but some are readable from further away, allowing data (e.g., card number and expiry date) to be slurped. Given the number of transactions made annually (more than 300 million in 2014),[31] illicit wireless access to card data is attractive, especially if merchants do not verify cardholder name and address information.[32]

## Takeaways

One should consider the benefits and risks of attaching devices to the Internet. (Why attach an Internet-enabled refrigerator to the Internet if one has to input the items place into and taken out of the refrigerator for the software to work properly? Do the devices need to be continually on the network or only on when a software patch is needed?) Securing the IoT is a business opportunity for home security companies to expand their service menu: ensuring the IoT home devices connect to a WAP using a security protocol, creating white lists (permitted connections) and black lists (prohibited connections) for IoT devices and Internet services, possibly keeping track of communications and sessions between devices and the Internet to look for permitted and "rogue" sessions, and so on. This will require upgrading current "minimally-secure" home routers and firewalls to "industrial strength" routers and firewalls.

With the adoption of IoT for commercial spaces—in smart buildings, for example—making mischief through WAPs moves from the individual to the wholesale level. The broadly publicized compromise of retailer Target's consumer data, accessed through a networked pathway between an HVAC service provider and Target corporate systems, illustrated dramatically the importance of supply chain security. Chapter 5, WAPs in Commercial and Industrial Contexts, explores WAP hacking opportunities in commercial and industrial spaces.

# ENDNOTES


1. Kacy Zurkus (14 January 2016), "Why Thinking Like a Criminal is Good for Security," *CSO*. Retrieved from http://www.csoonline.com/article/3022258/security/why-thinking-like-a-criminal-is-good-for-security.html?token=%23tk.CSONLE_nlt_cso_update_2016-01-14&idg_eid=fa2777600911eb0d655bcd255957a569&utm_source=Sailthru&utm_medium=email&utm_campaign=CSO%20Update%202016-01-14&utm_term=cso_update#tk.CSO_nlt_cso_update_2016-01-14
2. Raj Samani and Charles McFarland (2015), "Hacking the Human Operating System," *McAfee, Part of Intel Security*. Retrieved from http://www.mcafee.com/us/resources/reports/rp-hacking-human-os.pdf
3. Samani and McFarland, pp. 10–13.

4. Mathew J. Schwartz (28 July 2011), "RSA SecurID Breach Cost $66 Million," *Dark Reading*. Retrieved from http://www.darkreading.com/attacks-and-breaches/rsa-securid-breach-cost-$66-million/d/d-id/1099232? According to the article, the $66 million just covers the period between April and June 2011 when RSA was dealing with the fallout from the hack. Expenditures by EMC made during March 2011 to launch the investigation, harden systems, and counsel key customers are not included in this figure.
5. Arik Hesseldahl (4 April 2011), "RSA Explains How It Was Hacked," *All Things D*. Retrieved from http://allthingsd.com/20110404/rsa-explains-how-it-was-hacked/
6. Mat Honan (6 August 2012), "How Apple and Amazon Security Flaws Led to My Epic Hacking," *Wired*. Retrieved from http://www.wired.com/gadgetlab/2012/08/apple-amazon-mat-honan-hacking/all/
7. Mat Honan (17 August 2012), "How I Resurrected My Digital Life After an Epic Hacking," *Wired*. Retrieved from http://www.wired.com/2012/08/mat-honan-data-recovery/
8. Using a fake but credible credit card is also useful for vendors that insist on capturing credit card information even though you are signing up for free services. You can provide credit card information, it just will not be useable (or hackable)!
9. The supposed protection of only printing the last four digits of a credit card number is more apparent than real. Anyone taking a pizza order over the phone can see these last four digits, for example! (And if that order is for delivery, the home address is available, also). This false sense of safety that the truncated credit card number will not be useful to cyber criminals presumably encouraged those designing the Ashley Madison data confidentiality policies and practices to also store the last four digits of subscribers' credit card numbers (along with names and addresses) on the organization's servers in plaintext. See Kim Zetter (21 August 2015), "Answers to Your Burning Questions About the Ashley Madison Hack," *Wired*. Retrieved from http://www.wired.com/2015/08/ashley-madison-hack-everything-you-need-to-know-your-questions-explained/
10. Mat Honan (12 August 2012), "How I Resurrected My Digital Life After an Epic Hacking," *Wired*. Retrieved from http://www.wired.com/2012/08/mat-honan-data-recovery/
11. The US card issuers referenced are MasterCard, Visa, Discover, American Express.
12. Pablo's is an actual coffee shop in Denver (CO) that roasts its own Danger Monkey brand. Pablo's does not, however, make Wi-Fi available in its coffee shops. Patrons are thus chattier than the devices they don't bother plugging in! Theoretically, someone could, however, create a rogue hotspot to attract those unaware of the "no Wi-Fi" policy. Because it would not require hijacking the shop's communication signal—a cell phone would be sufficient—this would not be illegal.
13. Fahmida Y. Rashid (28 August 2015), "The Best VPN Services for 2016," *PC Magazine*. Retrieved from http://www.pcmag.com/article2/0,2817,2403388,00.asp
14. Kaspersky Labs' Global Research and Analysis Team (November 2014, Version 1.1), "The Darkhotel APT: A Story of Unusual Hospitality." Retrieved from https://securelist.com/files/2014/11/darkhotel_kl_07.11.pdf
15. Dan Goodin (20 October 2015), "Breaking 512-Bit RSA With Amazon EC2 Is a Cinch. So Why All the Weak Keys?" *ArsTechnica*. Retrieved from http://arstechnica.com/security/2015/10/breaking-512-bit-rsa-with-amazon-ec2-is-a-cinch-so-why-all-the-weak-keys/
16. Luke Valenta, Shaanan Cohney, Alex Liao, Joshua Fried, Satya Bodduluri, and Nadia Heninger (October 2015), "Factoring as a Service." *International Association for Cryptographic Research (IACR)*. Retrieved from https://eprint.iacr.org/2015/1000.pdf: Research funded by NSF Grant CNS-1408734, with additional funding from Cisco and an AWS Research Education Grant.
17. Claudia Buck (18 May 2014), "Microchips in Our Passports and Credit Cards: Are They Safe?" *The Sacramento Bee*. Retrieved from http://www.sacbee.com/news/business/personal-finance/claudia-buck/article2599038.html
18. Mark Roberti (16 August 2015), "The Internet of Hacked Things," *RFID Journal*. Retrieved from https://www.rfidjournal.com/articles/view?13372
19. Inspector Gadget was first popularized in the eponymous French-Canadian-American cartoon television series in the 1980s. The character later starred in a film, a 2015 TV series, and a video game.

20. Arjun Kharpal (21 February 2014), "Can Your Fridge Be Hacked in the 'Internet of Things'?" *CNBC*. Retrieved from http://www.cnbc.com/mobile-world-congress-event/
21. Darlene Storm (22 April 2015), "2 More Wireless Baby Monitors Hacked: Hackers Remotely Spied on Babies and Parents," *Computer World*. Retrieved from http://www.computerworld.com/article/2913356/cybercrime-hacking/2-more-wireless-baby-monitors-hacked-hackers-remotely-spied-on-babies-and-parents.html
22. Yael Grauer (5 September 2015), "Security News This Week: Turns Out Baby Monitors Are Wildly Easy to Hack," *Wired*. Retrieved from http://www.wired.com/2015/09/security-news-week-turns-baby-monitors-wildly-easy-hack/
23. Peter Sayer (2 February 2016), "Fitness Trackers Are Leaking Lots of Your Data, Study Finds," *CIO*. Retrieved from http://www.cio.com/article/3029153/fitness-trackers-are-leaking-lots-of-your-data-study-finds.html?token=%23tk.CIONLE_nlt_cio_mobile_2016-02-03&idg_eid=c9b38e5d8aee79f98d7ab9cca0e644d9&utm_source=Sailthru&utm_medium=email&utm_campaign=CIO%20Mobile%202016-02-03&utm_term=cio_mobile#tk.CIO_nlt_cio_mobile_2016-02-03
24. Snopes (11 November 2014), "Package Delivery Virus." Retrieved from http://www.snopes.com/computer/virus/ups.asp
25. Better Business Bureau (23 June 2015), "Summer Brings Slew of Shady Postcard Scams." Retrieved from http://www.bbb.org/wisconsin/news-events/consumer-tips/2015/06/summer-brings-slew-of-shady-postcard-scams/
26. See, for example, "Real Credit Card Numbers Credit Card Hack 2016." Retrieved from https://www.youtube.com/watch?v=mtCKiGsmN74
27. See, for example, Latest Hacking Software's, "Free Credit Card Numbers Generator 2016." Retrieved from http://www.latesthackingsoftware.com/free-credit-card-numbers-generator-2016/
28. Max Eddy (10 August 2015), "Don't Want Your Credit Card Hacked? Use Apple Pay," *PC*. Retrieved from http://www.pcmag.com/article2/0,2817,2489330,00.asp
29. Andrew Hudson (27 February 2015), "Student 'Hacks' Campus Card for Final Project," *CR80News*. Retrieved from http://www.cr80news.com/news-item/student-hacks-campus-card-for-final-project/
30. Mads Haahr (n. d.), "Introduction to Randomness and Random Numbers," *Random.org*. Retrieved from https://www.random.org/randomness/
31. Lisa Bachelor (23 July 2015), "Contactless Card Fraud Is Too Easy Says Which?" *The Guardian*. http://www.theguardian.com/money/2015/jul/23/contactless-card-is-too-easy-says-which
32. John Leyden (23 July 2015), "Contactless Card Fraud? Easy. All You Need Is an Off-The-Shelf Scanner," *The Register*. Retrieved from http://www.theregister.co.uk/2015/07/23/contactless_card_fraud_risk_which/

CHAPTER 5

# WAPs in Commercial and Industrial Contexts

## OVERVIEW

Smart is the new black for marketing technology applications: It goes with everything, whether applied to cars, shoes, mattresses,[1] or structures. This chapter looks at WAPs in built space that is not intended for private, residential use, but for industrial and commercial uses. Industrial facilities—classified here as environments in which the majority of the space is dedicated to the protection and care of machinery (e.g., manufacturing or food processing plants, power-generating facilities)—have a long history of embedded technology to control mechanical processes. Commercial spaces, on the other hand, are classified here as those environments in which the majority of the space is dedicated to sheltering products and people engaged in wakeful activities (e.g., retail stores and shopping centers, sports arenas and other entertainment areas, transportation depots, office complexes). Enabling the drive to make these venues smarter—a frequent euphemism for remotely controlled and programmable—wireless installations for connecting monitoring and sensing devices reduce the cost and disruption associated with upgrading hardwired cable plants. In addition to interrupting the point-to-point connectivity between devices, remote access to devices is effected by entrusting commands to IP connections for which the security of these mediated connections between devices and architecture was not initially designed. A gap is created between signal transmitter and signal receiver. That gap is the playground for enterprising hackers.

Being smart is not synonymous with having good judgment.

## OPPORTUNITIES

When making the calculation about which target will derive the highest yield, "A device that is both connected to the Internet and enables third-party



Hacking Wireless Access Points. DOI: http://dx.doi.org/10.1016/B978-0-12-805315-7.00005-X

remote access is an external attacker's prized desire."[2] The cost to the attacker of attempting digital access to a structure or asset—calculated as the risks of being caught or identified, wounded or killed by a security guard, prosecuted legally, incarcerated or otherwise penalized—is considerably lower than attempting simple brute force or even socially engineered physical access (e.g., tailgating, ID spoofing, distraction burglary). Granted, the digital hacker likely has more financial investment in tools and training. The immediate negative consequences of attempting a break-and-entry of a structure by using digital crowbars against wireless restraints are minimal, however, and physical access can be abetted by digital reconnaissance that enables disruption of locking mechanisms or surveillance and alarm equipment. Even improvements in digital forensics tools' techniques are neutralized by improvements in hacker tools that effectively mask or erase evidence of the intrusion attempt. Lag time between incident and detection—and between detection and response—is frequently lamented by security professionals and appreciated by hackers. Evidence includes high-profile cases like Sony, RSA, Target, JP Morgan Chase, Home Depot, and others.

Hacks leading to data compromise display a variety of incident patterns, as captured in Table 5.1, which shows results from analyzing three years of large data breach incidents investigated by Verizon and others who contributed

**Table 5.1** Three Years of Attacks Leading to Data Breaches[a]

| Incident Pattern ► / Industry (NAICS #) ▼ | POS Intrusions | WebApp Attacks | Cyber-espionage | Crimeware | Insider and Privilege Misures | Payment card Skimmers |
|---|---|---|---|---|---|---|
| Accommodation (72) | 53% | 1% | | | 3% | |
| Administrative (56) | | 4% | 1% | | 6% | |
| Educational services (61) | | 9% | 12% | 22% | 11% | |
| Entertainment (71) | 58% | 11% | 11% | | 5% | |
| Financial services (52) | | 17% | 1% | 21% | 6% | 7% |
| Healthcare (62) | 7% | 8% | 3% | 3% | 20% | |
| Information (51) | | 26% | 9% | 46% | 1% | |
| Manufacturing (31–33) | | 3% | 36% | 19% | 3% | |
| Mining (21) | | | 11% | | 67% | 6% |
| Other services (81) | 1% | 28% | 3% | 36% | 6% | |
| Professional services (54) | 2% | 2% | 26% | 10% | 1% | |
| Public (92) | | | | 18% | 26% | |
| Retail (44–45) | 20% | 2% | | 25% | 1% | 4% |
| Transportation (48–49) | | | 41% | 9% | 18% | 5% |
| Utilities (22) | | 17% | 50% | 17% | | |

Legend: < 10% | 11-19% | > 20%

[a]*Verizon Enterprise Solutions (March 2016)*, Data Breach Digest: Scenarios from the Field. *Retrieved from http://www.verizonenterprise.com/resources/reports/rp_data-breach-digest_xg_en.pdf.*

field reports for inclusion in the data digest. The results are sorted by primary North American Industry Classification Standard (NAICS) codes of the compromised organizations. Cyber-espionage figures especially prominently in critical infrastructure sectors like manufacturing, transportation, and utilities. These are also industry sectors that share certain technology platform characteristics, which will be discussed in the section on industrial applications.

## COMMERCIAL SPACE TAXONOMY

All commercial spaces are not created equal with respect to how assets are valued by facility owners, managers, stakeholders, and potential hackers. These spaces also vary widely in terms of purpose, traffic, layout, and technology. They generally cluster along the "Four Ps" of access control preferences—promiscuous, permissive, prudent, paranoid—with respect to how physical access into the area is controlled. Existing technology implementations do not always align with and enable these declared access control preferences, however, and can create a serious disconnect between the level of physical and digital access control desired and the level actually achieved. This is especially evident in industry sectors that still rely on technology platforms built for point-to-point, direct connectivity (rather than the networked connectivity). Table 5.2 shows a general taxonomy of commercial spaces cross-referenced by physical access control philosophy and digital access control platform vulnerability. This will set the stage for discussion of attack scenarios based on hacking WAPs used in industrial and commercial applications.

Using shopping centers and malls as an example, one can see that implementing sophisticated technology, even technology explicitly designed for Internet connectivity (as opposed to legacy systems designed for point-to-point connectivity), can open gaping holes in security systems and even abet attacks on remote targets if the technology is not deployed with the understanding that connected devices are WAPs. Intentionally promiscuous with respect to allowing anyone to enter physically, shopping malls are investing increasingly in sophisticated surveillance technology and training for security officers and retail/concessionaire employees. The International Council of Shopping Centers has invested $2 million to develop courses for security officers. A large (one million square foot (SF)) shopping center will invest up to that amount in annual security measures that include hundreds of surveillance cameras.[3] As often is the case, however, that security investment can be undermined when insufficient attention is paid to the WAP nature of those cameras, which are part of a small computer system. Failure to harden cameras with robust login credentials (e.g., not changing manufacturer's default passwords) or restrictions on connection requests enables botnet recruitment and malware downloading.[4] This can work to the advantage of hackers

**Table 5.2** Comparison of Representative Commercial Venues, Physical Access Control Philosophy, and Digital Access Control Vulnerability

| Representative Venues | Physical Access Control Philosophy | | | | Digital Access Control Vulnerability | | | |
|---|---|---|---|---|---|---|---|---|
| | Paranoid | Prudent | Permissive | Promiscuous | Paranoid | Prudent | Permissive | Promiscuous |
| Church, synagogue | | | | X | | | X | X |
| Concert hall lobby | | | | X | | | | X |
| Gym | | X | X | | | | | X |
| K12 school | X | X | | | | X | X | |
| Manufacturing facility | | X | X | | | X | X | |
| Medical care | | X | X | | | X | | |
| Office building (government) | X | X | X | X | X | X | | |
| Office building (private) | | X | X | X | | X | | |
| Prison, jail | X | | | | | X | | |
| Research laboratory | X | X | | | | X | | |
| Retail shop | | | | X | | | X | X |
| Shopping mall | | | | X | | | X | X |
| Sports arena | | | X | X | | | | X |
| Transportation lobby | | | | X | | | | X |
| University/college classroom | | | | X | | | X | X |
| University/college residence | | X | X | X | | X | X | X |
| Utility company | | X | X | | | | X | X |

looking to launch a distributed denial of service (DDoS) attack by exploiting Shellshock vulnerabilities[5] and still-enabled Telnet ports.[6]

## Industrial Applications

ICS encompass a range of systems and connected devices (e.g., SCADA systems; distributed control systems; and programmable logic controllers, PLCs). They are "combinations of control components (e.g., electrical, mechanical, hydraulic, pneumatic) that act together to achieve an industrial objective (e.g., manufacturing, transportation of matter or energy)."[7] They are frequently systems that were intended for use within a local, confined physical area and not over shared communications channels: Point-to-point connectivity or direct physical access was required to make configuration changes. Similarly, these systems typically cannot be upgraded or patched from a central location. The systems are designed as long-term solutions with very limited downtime. As a consequence, fixes are applied irregularly and legacy systems are not assigned an end-of-service date as are components of an IT system.

Control systems can be classified according to whether they are intended for manufacturing use case scenarios or distribution. The former features more closely defined physical boundaries and direct communications links (i.e., wired vs. wireless, although that is changing with improvements in wireless reliability and throughput), whereas the latter often features broad geographic dispersion and switched or wireless communications links. Because signal latency can hinder the effectiveness of either category of control systems, some security practices recommended for hardening an IT system are contraindicated for an ICS.

The security triad of CIA still applies, but unlike most applications in which confidentiality is the most valued objective, here the continued availability of systems is the primary objective, followed by integrity and then confidentiality.

A partial list of attendees (Fig. 5.1) at the annual ICS cyber security conference, first held in 2002, hints at the variety of industry sectors that rely on ICS and are concerned about their vulnerabilities.

## Burrow Attack: Stuxnet (Industrial Sabotage)

*Hacker Objective:* Command and control (C&C) over mechanical systems (most famously, an Iranian nuclear reactor)

*Hacker Technique:* Multilayer, multiple techniques (social engineering, credential compromise, four zero-day exploits)

*Victim Impact:* Physical damage to centrifuges; disruption of processes dependent on centrifuges (used to separate uranium isotopes: U-235 from U-238)[9]

Stuxnet was the big wakeup call that proved, even to skeptics, that the postulated disruption of mechanical systems—demonstrated in 2007 with the cyber-kinetic takedown of an industrial turbine by Idaho National Laboratory researchers—was feasible. The staged attack in 2007 prompted both mass media publicity and cautions against hype. The acting undersecretary of DHS's National Protection and Programs Directorate at the time stated, "several conditions have to be in place. ... You first have to gain access to that individual control system. [It] has to be a control system that is vulnerable to this type of attack. ... It is a serious concern. But I want to point out that there

**FIGURE 5.1**
Sample of industry participants at ICS cyber security conferences.[8] *Figure used with permission from the SecurityWeek ICS cyber security conference 2016 website. Retrieved from http://www.icscybersecurityconference.com*

is no threat, there is no indication that anybody is trying to take advantage of this individual vulnerability."[10]

The (most likely) nation-state-produced, but not staged, attack against an Iranian nuclear reactor in 2009 demonstrated irrefutably that a well-funded, determined hacker team with insider knowledge about the Siemens s7 PLC could successfully compromise even hardened ICS. (A PLC vulnerability enabled infection spread via a compromised USB drive.) The attack against the Natanz uranium enrichment site was just one of the state-sponsored attacks in Iran.[11] Under development since at least 2005, the actual Stuxnet attack against Iran's Natanz site combined social engineering, MITM attack of the PLC (through compromised credentials), and four zero-day exploits.

### *Victim's Chronology*[12]

1. An infected USB drive is inserted into a computer on the Natanz network, whether intentionally (deliberate insider threat) or accidentally (e.g., the "dropped USB" ploy discussed earlier). The worm then uploaded to the facility's network.
2. Stuxnet worm follows its programmed code to discover system-controlling software (Siemens s7).
3. Stuxnet takes over control of the centrifuges under the software's control and either reprograms valves to close (version 0.5) or to alternate centrifuge cycle speeds (version 1.x), thereby accelerating centrifuges until spinning dangerously, then decelerating over a series of months.[13]
4. Weakened centrifuges had to be replaced: approximately 20%.

Just as attacks on biological systems wreak particular havoc when they cross species, so do attacks on cyber-kinetic-mechanical systems. The Stuxnet worm, for example, traveled to an estimated 100,000 computers, most of which were not associated with the initial target, Iran's nuclear power assets. Different versions are at play. Although the original version (0.5) did not exploit Microsoft vulnerabilities and compromise centrifuge speed, later versions (1.x) did. Version 0.5 spread through removable media and email attachments. It contacted four C&C servers, hosted by commercial providers, which appeared to belong to an Internet advertising agency: Media Suffix. Version 1.x is more aggressive and created more collateral damage.[14]

Other attacks against industrial systems, attacks like Night Dragon (data theft)[15] and Shamoon (sabotage through the destruction of data)[16] that have targeted assets in the oil and gas industry (e.g., Saudi Aramco in 2012), reveal the persistent, methodical, multilayered techniques characteristic of state-sponsored activity. WAP exploitation provides an additional potential entry point for these malware products. It can accelerate their proliferation through the immediately targeted system and contribute to collateral damage that occurs when related systems are infected. Common attack surfaces are compromised individual mobile laptops (via spear phishing), although other common "business productivity" devices can also be compromised. The use of under-protected, off-the-shelf, consumer-grade applications—in conjunction with ICS environments that have historically been kept in isolation—is as potentially dangerous as sharing kindergarten toys would be for someone lacking effective immune systems ("boy in the bubble" syndrome).

At the individual consumer level, we are highly dependent on, but largely ignorant of, these ICS environments that represent critical infrastructure sectors, in particular, the chemical, dams, transportation systems, critical manufacturing, energy, nuclear reactors (and materials and waste), and water and wastewater sectors. Some medical equipment is susceptible to the kind of sabotage wreaked by Shamoon and Night Dragon when data or programmed code is either overwritten or intercepted. With the advent of smart cities, transportation systems, and even individual vehicles, the challenge of securing wireless devices from malicious attack becomes more urgent.

## Commercial Applications

Other scenarios for using automated system are not as closely associated with the control of heavy-duty machinery and mechanical processes. Found in use cases like digital locks, sensors, environmental controls, and surveillance/physical security equipment, these commercial applications are frequently managed through WAPs. Smart buildings fall under this category.

Being more responsible and sustainable—more green—is frequently touted as the best justification for investing in smart technology for regulating building energy use, as is cost savings. But what is the total cost of ownership if security is not built into systems before deployment? Again, what is convenient for facilities managers is also convenient for hackers looking for a vulnerable WAP as an entry point. Indeed, the dilemma for the hacker might be to choose just one. Cisco's openBerlin Innovation Center illustrates just how many sensors it can take to automate HVAC and illumination and eliminate manual switches: About 10,000 sensors are estimated for the 13,000 SF building.[17] During regular work hours that equates to about 100 sensors per employee; the ratio changes, of course, nights, weekends, and holidays.

Again, in the tradeoff between cost and security, the latter can lose out. Consider the importance of centrally controlled, digital locks in maximum security prisons, for example, as a compensating control for the lopsided ratio of prisoners to guards. The central locking systems have cascading release functions, that is, they "fail open" in case of fire or other such emergency. One Florida prison experienced a couple of seemingly spontaneous "cascading releases" in its maximum security wing in 2013 that set all prisoners there free.[18] Whether a malfunction, internal collusion, or external hack the concern is nonetheless vivid.

The locking systems implemented in prisons are often not sufficiently hardened. Usually the PLCs that control mechanisms like doors are programmed using Ladder Logic, a simple program that is convenient for programmers—and vulnerable to exploitation. Prepackaged exploits are available online at <Exploit-DB.com>.[19] PLCs reduce wiring costs (as do wireless networks) and are desirable in these early examples of smart buildings that use electronic systems to control intercoms, illumination, surveillance, water, bathing facilities, and so on. Good news for the would-be hacker: "Access to any part, such as a remote intercom station, might provide access to all parts."[19]

Due to the "always on" requirement for PLCs that control the physical world (e.g., door locks, environmental controls, surveillance), deploying an intrusion prevention system (IPS) is not desirable. The IPS is built to interrupt processes if an anomaly is detected. Consider, for example, the fail-open design of building egress or even prison cell doors. Although the PLC default position might be locked, during a power failure or emergency doors are programmed to open so that people are not trapped. Downtime for maintenance thus is closely coordinated and minimized. The weekly patching schedule common in an IT system (in contrast to an IC system) would not be practical.

These PLC environments cannot be disconnected from the Internet without disrupting business activities. In the case of prisons, for example, both operations (e.g., commissary replenishment and other supply chain coordination activities) and administration (e.g., mandatory reporting to federal oversight agencies) require connectivity to the outside world. Wireless channels are used for improved surveillance coverage, reduced network cabling costs, and for mobile applications. Patrol cars covering a facility's perimeter, for example, transmit videos wirelessly and unencrypted to guards inside. Those and other videos are easy to find on YouTube and various other mass media websites.[19]

Automated sensors like motion- and heat-detection devices are also used for perimeter control in built environments ranging from residential neighborhoods to skyscrapers to civilian and military government facilities to transportation depots to power-generating stations. High-value, albeit geographically remote and minimally manned, assets like power transmission stations present an attractive use case scenario for deploying sensors to detect physical intrusions, geophysical disturbances (e.g., earth tremors), utility equipment condition, and environmental hazards (e.g., fire). Using ultra low-power, tiny sensors, which are immune to high voltage, in a mesh network can offer coverage over a broad area surrounding transmission stations, eliminating the need for on-site personnel and fixed equipment installation like cameras.[20] The security challenge is not trivial, however. Requirements for low battery use compete with the energy drain presented by the more intensive processing power needed for encryption or an IPS (to protect the digital information). Each solution seems to beget another potential vulnerability.

In addition to building access control, other uses for ICS in any smart building (not just a personnel containment facility) underscore the dependence on consistent power supply and automated controls for which availability—continuous functioning—overshadows latent concerns about confidentiality and integrity. Smart buildings act as a microgrid; with increased deployment of solar and other alternative sources of energy, this characteristic will become even more prevalent. Wireless sensors are used to help control energy use; detect smoke and fire; monitor HVAC systems (essential to maintaining temperature control for heat-averse servers and other computing equipment) for fault detection, diagnosis, and prognosis; promote worker safety and reduce injury in manufacturing facilities. Other sensors are used to support human comfort and hygiene needs, for example, lighting and touch-free water faucets, soap dispensers, and hand dryers.

## Takeaways

What is convenient for one is also often convenient for another. Likewise, a cost savings for one can also mean a cost savings to another. Relying on a uniformly digital strategy, even one that uses an array of products for protecting various OSI layers, is less diverse than one that integrates analog along with digital products.[21] Raising a hacker's cost of exploitation by layering digital and analog control mechanisms can reduce your attractiveness as a target; the risk-averse, profit-motivated hacker may move on. (All bets are off, however, for the state-sponsored hacker.)

Depending on technology alone as the hardening guarantor for any system is a disingenuous way of outsourcing responsibility and falls short of meeting due care and due diligence commitments. Facilities managers may hope to simplify their job responsibilities and costs by avoiding the vagaries of recruiting, managing, training, and retaining people, and accommodating occasional personnel "outages" (e.g., holidays, weekends, sickness, resignation/termination). Automated systems can be set up as "always on" (given sufficient redundancy) with reduced expectation of malfunction. Still, the possibility of cascading failures across systems that are all interconnected digitally requires their hardening and, perhaps, their digital discontinuity through the implementation of analog components as additional security gates. If the legitimate remote administrator cannot move laterally across networks, neither can others (as happened to Target).

*What does a hacker not want to encounter?*

- Isolated, hardened credentials that are stored prudently (no password files entitled *passwords* or *access codes*, etc.)
- Unique administrator accounts that do not use manufacturer defaults
- Mechanical systems that require physical access
- Security that lives with data under all three states—at rest, in motion, or in process—as well as at the edge of communications networks
- Encrypted endpoint devices

NIST has compiled a solid list of guidelines for addressing security in automated systems. These guidelines should apply at all layers of the embedded architecture including WAPs. The latter must be hardened and protected as well. The NIST guidelines recommend the following security objectives:

- Restrict logical access to the ICS network and network activity
- Restrict physical access to the ICS network and devices
- Protect individual ICS components from exploitation
- Restrict unauthorized modification of data
- Detect security incidents and incidents
- Maintain functionality during adverse conditions
- Restore the system after an incident[22]

Implementing these objectives successfully requires thoughtful deployment of technology, policies, and personnel. User activity with respect to devices, including WAPs, should be tracked by individuals rather than by user role accounts (e.g., system administrator). The business-to-business or machine-to-machine activity must also be tracked and parameters for normal behavior set to facilitate anomaly detection.

Segregation of ICS from internal administrative networks and both from public access networks within facilities should be seriously considered. The FAA, for example, has mandated that on-board Internet access on aircraft have no common components with the communications infrastructure used to control and communicate with the aircraft. This is to protect against hackers taking down an airplane. Such a best practice should be adopted by ICS users since the cost of multiple wireless systems in a facility (each with its own unique encryption passcode) can be much less than the cost of recovery after a breach (approximately $179 per record compromised[23]).

Connectivity between ICS, administrative networks, and public access networks should be tightly controlled and monitored. Gateways between the networks should be limited and provisioned with policies to restrict the flow of data. Anomalies should be routed and handled by personnel to ensure the communications are appropriate. Security should not be relaxed to support a business partner. These best practices have been successfully used by the US Department of Defense (DoD); businesses need to adopt these practices in their ICS environments. Approximately 82% of critical infrastructure that is dependent on ICS is owned and managed by private sector, nongovernmental organizations.

Remote communications into ICS and administrative networks should be secured through the use of VPNs, and possibly additional encryption mechanisms, because nonsecure communications over the Internet and over WAPs make industrial espionage easier. When designing the wireless network architecture, latency issues must be considered: Mission-critical controls on ICS require a dedicated network without encryption. Monitoring the ICS remotely does not require subsection response time. Here one should use encryption with the remote monitoring (via a gateway, not any ICS device) so hackers cannot see this traffic.

These three features (network separation of functions, strong controls of information flows across networks, and secure remote communications) will make organizations and companies a less desirable target of convenience. For critical infrastructures and critical manufacturing and processing (e.g., food and medical) organizations, these features are necessary but not sufficient to protect against industrial espionage and sabotage. Here intrusion detection systems (IDS) and analysis of device logs (e.g., database access, performance and utilization logs for robotics) for anomalies can act as an early warning system for attempted and actual breaches of networks. An IDS is preferable to an IPS because the latter can be gamed by a hacker to shut down the system. While it is likely APT will most likely occur on one's public network, monitoring of all networks is needed to ensure that malware is not (intentionally or unintentionally) introduced into the administrative and ICS networks.

Automated business-to-business activities generate a significant amount of IoT value: as much as 70% according to one analyst group.[24] Individual, robust (not manufacturer's default) passwords are essential, as is separation of duties on the part of administrators wherever possible. External access by vendors should be limited and segregated from other network access—and vendor internal security policies should align with those of customers so that neither becomes a target (or Target). The convergence of the physical and digital environments presents challenges when disparate security priorities—confidentiality, integrity, security—compete for precedence within the same organization. The health care industry combines those environmental challenges with critical, life-and-death consequences when a WAP failure in either the IT or IC systems bleeds into the other.

## ENDNOTES


1. Stephanie Mlot (19 April 2016), "Is Your Partner Cheating? This Mattress Will Alert You," *PC Magazine*. Retrieved from http://www.pcmag.com/news/343828/is-your-partner-cheating-this-smart-mattress-will-alert-you.
2. Idan Tendor (12 October 2015), "Top 5 Security Threats from 3rd Parties," *Network World*. Retrieved from http://www.networkworld.com/article/2991914/network-security/top-5-security-threats-from-3rd-parties.html.
3. Ronda Kaysen (26 November 2013), "Malls Work on Their Security, but Keep it in the Background," *The New York Times*. Retrieved from http://www.nytimes.com/2013/11/27/realestate/commercial/malls-work-on-their-security-but-keep-it-in-the-background.html?_r=0.
4. Graham Cluley (27 October 2015), "Hacked Shopping Mall CCTV Cameras Are Launching DDoS Attacks," *Tripwire*. Retrieved from http://www.tripwire.com/state-of-security/security-data-protection/hacked-shopping-mall-cctv-cameras-are-launching-ddos-attacks/.
5. Shane Shick (18 November 2014), "Latest Shellshock Attack Uses Bashlite to Target Devices Running BusyBox," *Security Intelligence*. Retrieved from https://securityintelligence.com/news/latest-shellshock-attack-uses-bashlite-target-devices-running-busybox/.

6. Lucian Constantin (17 November 2014), "ShellShock Exploiting Bash Malware Targets Embedded Devices Running BusyBox," *PCWorld*. Retrieved from http://www.pcworld.com/article/2848692/bash-malware-targets-embedded-devices-running-busybox.html.
7. Keith Stouffer, Victoria Pillitteri, Suzanne Lightman, Marshall Abrams, and Adam Hahn (May 2015), "Guide to Industrial Control Systems (ICS) Security," NIST Special Publication 800-82, Rev. 2. Retrieved from http://nvlpubs.nist.gov/nistpubs/SpecialPublications/NIST.SP.800-82r2.pdf.
8. Figure used with permission from the ICS Cyber Security Conference 2016 Website. Retrieved from http://www.icscybersecurityconference.com/#!about/c4nz [PERMISSION NEEDED].
9. Nuclear centrifuges are sophisticated, computer-controlled instruments that rotate at some 100,000 rpm. Separation of the uranium isotopes is an intermediate step in the process of producing enriched uranium metal for use as fuel for nuclear power generation or for a nuclear bomb. Marshall Brain, "What's a Uranium Centrifuge?" *How Stuff Works Website*. Retrieved from http://science.howstuffworks.com/uranium-centrifuge.htm.
10. Jean Meserve (26 September 2007), "Staged Cyber Attack Reveals Vulnerability in Power Grid," *CNN*. Retrieved from http://www.cnn.com/2007/US/09/26/power.at.risk/index.html?iref=topnews.
11. Dan Goodin (16 February 2016), "Massive US-Planned Cyberattack against Iran Went Well Beyond Stuxnet," *ArsTechnica*. Retrieved from http://arstechnica.com/tech-policy/2016/02/massive-us-planned-cyberattack-against-iran-went-well-beyond-stuxnet/.
12. BBC, "Timeline: How Stuxnet Attacked a Nuclear Plant," *BBC*. Retrieved from http://www.bbc.co.uk/timelines/zc6fbk7.
13. Geoff McDonald, Liam O. Murchu, Stephen Doherty, and Eric Chien (26 February 2013), "Stuxnet 0.5: The Missing Link," *Symantec*. Retrieved from http://www.symantec.com/content/en/us/enterprise/media/security_response/whitepapers/stuxnet_0_5_the_missing_link.pdf.
14. Dan Goodin (26 February 201e), "Revealed: Stuxnet 'Beta's' Devious Alternate Attack on Iran Nuke Program," *ArsTechnica*. Retrieved from http://arstechnica.com/security/2013/02/new-version-of-stuxnet-sheds-light-on-iran-targeting-cyberweapon/.
15. Among others, see McAfee Foundstone Professionals and McAfee Labs (10 February 2011), "Global Energy Cyberattacks: 'Night Dragon'." Retrieved from http://www.mcafee.com/us/resources/white-papers/wp-global-energy-cyberattacks-night-dragon.pdf.
16. Christopher Bronk and Eneken Tikk-Ringas (1 February 2013), "Hack or Attack? Shamoon and the Evolution of Cyber Conflict," *Rice University*. Retrieved from http://bakerinstitute.org/media/files/Research/dd3345ce/ITP-pub-WorkingPaper-ShamoonCyberConflict-020113.pdf.
17. Mark Halper (22 February 2016), "No Light Switches at Cisco's Own Berlin Smart Building, Where You Can't Miss the IoT," *LEDs Magazine*. Retrieved from http://www.ledsmagazine.com/articles/2016/02/no-light-switches-at-cisco-s-own-berlin-smart-building-where-you-can-t-miss-the-iot.html.
18. Kim Zetter (16 August 2013), "Prison Computer 'Glitch' Blamed for Opening Cell Doors in Maximum-Security Wing," *Wired*. Retrieved from http://www.wired.com/2013/08/computer-prison-door-mishap/.
19. Teague Newman, Tiffany Rad, and John Strauchs (30 July 2011), "SCADA and PLC Vulnerabilities in Correctional Facilities," p. 12. Retrieved from http://www.wired.com/images_blogs/threatlevel/2011/07/PLC-White-Paper_Newman_Rad_Strauchs_July22_2011.pdf.
20. J. Roger Bowman and Darrin Wahl (June 2012), "Advanced Distributed Sensor Networks for Electric Utilities," *SAIC Report for California Energy Commission (CEC-500-2012-069)*. Retrieved from http://www.energy.ca.gov/2012publications/CEC-500-2012-069/CEC-500-2012-069.pdf.
21. For a discussion of "massive-scale software diversity," see: Per Larsen, Stefan Brunthaler, and Michael Franz (March 2014), "Security Through Diversity: Are We There Yet?" *IEEE Security and Privacy Magazine*, Vol. 12, No. 2, 28–35. Retrieved from https://www.researchgate.net/publication/261718805_Security_through_Diversity_Are_We_There_Yet.
22. NIST 800-82, p. 3.
23. Ponemon Institute (27 May 2015), "Cost of Data Breach Grows as Does Frequency of Attacks," *Ponemon Institute*. Retrieved from http://www.ponemon.org/blog/cost-of-data-breach-grows-as-does-frequency-of-attacks.

24. McKinsey report cited in Centri's white paper, "Securing the Internet of Things: 5 Key Areas to Protect," p. 4. Retrieved from http://www.centritechnology.com/securing-the-iot-white-paper/ (McKinsey Global Institute (June 2015), "The Internet of Things: Mapping the Value Beyond the Hype," Retrieved from http://www.mckinsey.com/~/media/McKinsey/dotcom/Insights/Business%20Technology/Unlocking%20the%20potential%20of%20the%20Internet%20of%20Things/Unlocking_the_potential_of_the_Internet_of_Things_Executive_summary.ashx).

CHAPTER 6

# WAPs in Medical Environments

Several forces combine to make wireless technology especially prevalent in medical facilities: cost of implementing or upgrading wired networks; inconvenience and environmental burden (construction dust, noise, repair work) of installing cable; tight patient exam scheduling and the desire for on-demand, real-time patient information (e.g., electronic health records or EHRs); and the use of sensors, medical devices, and building automation solutions. The importance of WAPs in medical practice is underscored by estimates about the size of the global wireless EHR market: about $11.2 billion in 2013 and projected to at least double that size in 2018, according to BCC Research. This market research group defines wireless medical market segments as products that include wireless patient monitoring, EHR-compatible devices, wireless EHR software, EHR mobile technologies, application markets, pediatric growth trackers, capnography,[1] real-time location devices, video applications, and patient/guest Internet access, in addition to end-user (consumer) markets.[2] Consumer products include such devices as fitness trackers and sleep monitors, as were discussed in Chapter 4, Hacks Against Individuals. This chapter looks at wireless devices used as integral components of formally designated medical environments; their role in promoting patient, clinician, organization, and facility well-being; their attack surface characteristics and attractiveness to hackers; and the challenge of balancing the security objectives of CIA.

The medical industry is where things get critical, cynical, and even more complex than in consumer, manufacturing, and commercial environments. There really can be blood on the floor when security goes awry: it's more than inconvenience and compromised confidentiality. Parallels can be drawn between physical/biological compromise in medical care facilities and digital/logical compromise. The spread of infection between biological systems is similar to the spread of infection between digital systems.

Hospitals are notorious breeding grounds for biological infection: "on any given day, approximately one in 25 US patients has at least one infection



Hacking Wireless Access Points. DOI: http://dx.doi.org/10.1016/B978-0-12-805315-7.00006-1

contracted during the course of their hospital care."[3] The death rate from such infections in 2011 was more than 10%. Health care-associated infections include those contracted through invasive procedures (central-line associated bloodstream infections or CLABSI, catheter-associated urinary tract infection or CAUTI, surgical site infection or SSI, and ventilator-associated pneumonia or VAP). Presurgery protocols call for taking antibiotics internally a few days prior to surgery, yet ambient bacteria can still cause problems (studies indicate that cleaning tools like mops and dust cloths can spread contamination) in spite of using tons of disinfectant products.[4] Of course, using the products is controversial because of the increasingly resistant bacterial strains that are implicated in life-threatening infections. As with digital infections, we are in a perpetual game of catch-up. It is not surprising that infection transmission thrives when sick people with already compromised immune systems and possibly open wounds are concentrated in one location. Likewise, it is not surprising that malware transmission thrives in technology environments in which disparate devices with already weak (or nonexistent) built-in security protection, lightweight access controls, and ineffective credential management are sharing a wireless network.

The precarious physical health of those in hospitals and other medical care facilities is analogous to what happens when computing technology is introduced without a complete understanding of the fundamental requirements for protecting data health (i.e., its CIA). The aggressive push for EHRs did not effectively prescribe how to prepare data for automated sharing. Data collection points are legion. Sensors abound, as do mobile devices, to receive and transmit data for storing, sharing, and further analysis. There is an inherent tension between the convenience of unimpeded wireless access to patient records, diagnostic information, and treatment results, and the extra steps required to ensure data and communication security (as represented by the CIA triad).

Medical staff have become accustomed to relying on quick retrieval of data to determine proper diagnosis and treatment for patients. Encrypting that data slows the process by adding overhead to processing time—but, with very high-speed networks (in the gigabit per second range), delays are measured in fractions of a second. Staff may choose to ignore encryption options, which require an extra authentication step via password or token, for sharing needed information like X-ray images and just opt for using unprotected messaging channels.[5]

Implementing multiple firewalls to separate data stores likewise slows the retrieval process as does segregating various device traffic on different network segments, adding complexity to traffic management and change control processes, and a risk to QoS requirements. According to some practitioners, such complexity is not readily scaleable.[6]

The health information technology (HIT) environment is one in which latency—lag time in communication signals between medical resources (human, machine, data)—can lead to injury, suboptimal outcomes, and even death. From a technical perspective, however, such latency concerns are only valid when there is a single network for all devices. Segregation of networks allows different latency and thus different security support. It is the machine-to-machine signals for which quality of service (QoS) must be maintained at the highest levels, for example, when performing robotic or remote surgery. Human-to-human or human-to-machine communications are more tolerant. The priority afforded availability as a security objective competes with integrity as the primary objective. Message integrity, confirmation that content and source are authentic and authoritative, is critical. Still, public policy concerns about—and legal consequences associated with—confidentiality of data have influenced security architecture and budgeting decisions. Availability and integrity trail confidentiality in terms of senior management decisions about how to prioritize security objectives and where to make security—and technology—investments.

This is really an area for further analysis to distinguish between concerns based on convenience factors versus system responsiveness (i.e., speed at which system responds to requests), impacts to efficiency (e.g., number of patients supported per resource or throughput), and effectiveness (e.g., patient mortality rate, patient readmission rate, or output/outcomes). If system responsiveness were critical to hospital effectiveness, one would expect to see noticeable differences when a hospital moves from primary systems to backup systems (e.g., paper operations). These differences would likely be reported to the government and then (intentionally or unintentionally) made available to the press. Analysis of normal systems and backup systems (due to hacks, power failures) for efficiency and effectiveness is an area for further analysis.

Health care environments add life-and-death concerns on top of the typical information concerns about CIA. Health Information Portability and Accountability Act of 1996 (HIPAA) is the legislation that imposes penalties on medical practitioners, facilities, and business associates when patient PII is compromised. Less visible among the high-profile incidents of health record data breaches are violations of data integrity and system availability. The successful ransomware attack on Hollywood Presbyterian Hospital in early 2016, in which a variety of hospital assets—not just patient records—were held hostage, provided irrefutable evidence that wireless vulnerabilities create opportunities for hackers.

## MEDICAL EHRs

### Confidentiality

EHRs provide a rich store of PII including more than 18 different identifiers (e.g., name, address, SSN, date of birth), in addition to payment information, medical conditions, and treatments.[7] This is more PII and private information than one's bank collects. And all that information is valued by hackers to ensure income flow, whether by selling it on the black market (health insurance credentials can be worth 20 times more than a credit card)[8] or by committing billing fraud. Such health care fraud "accounts for 3%–10% of annual U.S. health expenditures" or at least $74 billion a year.[9] The information is also used for filing fraudulent income tax refund claims.

HIPAA addresses data integrity in §164.304—Definitions:

> Integrity means the property that data or information have not been altered or destroyed in an unauthorized manner.

HIPAA was one of several pieces of legislation passed in the 1990s[10] that addressed citizen and politician concerns about ensuring the "right to be left alone."[11] The first civil monetary penalty for noncompliance by a covered entity (Cigna Health, in this case) was not imposed until 2011, however. This was after passage of the Health Information Technology and Economics Clinical Health (HITECH) Act, which defined higher fines for HIPAA violations.[12] Prior to this, Providence Health System was fined $100,000 for casual treatment of protected health information belonging to more than 386,000 patients)[13] and CVS was fined $2.25 million for improperly disposing of prescription labels and other identifying information in unsecured trash bins.[14]

Since 2005, the Privacy Clearinghouse has received more than 551 reports of breaches in the medical sector associated with portable devices, hacking or malware, and unknown vectors.[15] The amount of health PII accessed by hackers shows little sign of abating. One research firm predicts that "one out of three individuals will have their medical records compromised by cyberattacks in 2016."[16]

### Anthem Attack

*Hacker Objective:* Primary objective was to obtain information about key personnel in aerospace, energy, etc. as part of a suspected state-sponsored APT attack; the secondary objective was to obtain a rich information store for mischief

*Hacker Technique:* Multilayer, persistent, multiple techniques (watering hole,[17] credential compromise, zero-day exploits, Sakurel malware enabling backdoor)[18]

*Victim Impact:* Compromise of 80 million records containing PII (carries fines to insurance organization of about $200 per record,[19] although Anthem carried more than $100 million in data breach insurance); compromise of victim organization's family of businesses for another 22.1 million records compromised;[20] predicted increase in individual fraud cases.

As of May 2016, details describing the steps taken in the Anthem attack were not available because of the ongoing investigation. Publicly available information indicates that the credentials of at least five employees with privileged IT access were stolen. Common techniques include phishing email messages (e.g., Sony 2015 attack) and malware propagation that allows backdoor access into systems. These techniques are often associated with mobile computing and storage devices, in addition to intercepted signals from WAPs. Mitigating measures that were not in place at Anthem include record encryption (not required by HIPAA for dedicated servers), robust protection of access credentials, least privilege policy (firmly limiting privileged access), effective employee awareness training, and intrusion and anomaly detection.[21] Organizational inattention to security needs was also evident. Anthem was cited in 2013 by the US Department of Health and Human Services (HHS) and fined $1.7 million for not performing a complete risk analysis when a new, online customer portal was launched.[22]

From the attacker side, the apparent connection between different hacker groups—based on the timing and techniques used for attacks that include Anthem as well as the US Office of Personnel Management (OPM)—is disturbing. The malware used in both, Sakula, is a RAT with many different strains. The Deep Panda and Black Ivy groups are associated with this malware.[23]

## EHRs AND MEDICAL DEVICES

### Integrity

Although valued for their potential to facilitate information sharing horizontally across multiple care providers (internal and external to a specific medical facility) and longitudinally over years, EHRs are susceptible to easy alteration if safeguards are not implemented. It is essential that access control principles be applied consistently to ensure that information is reported by appropriate personnel (individually identifiable) and that information updates be time-sampled to support nonrepudiation and a kind of information "chain of custody." Likewise, change control principles must be adhered to so that the sequence of interventions and events can be understood easily and patient care can be monitored and adjusted as needed. Effective implementation of access and change/configuration control mechanisms means that all health information sources within the facility's ecosystem must be validated. Many of these information sources are wireless.

Clinician convenience and intensive scheduling demands can combine to create an environment in which data integrity cannot be ensured and receives only intermittent scrutiny. Record integrity is compromised, for example, when clinician notes are cloned (copied-and-pasted) from one patient to another or from one patient screening report to another, when dictation errors are accepted without validation, or when template documentation is inadequate for describing the patient condition. In addition to nonmalicious errors, inadequate data integrity can promote health care fraud and abuse.[24]

Wired connectivity is bolstered and augmented—even sometimes replaced—in medical facilities by the use of WAPs. This results in significant savings in plant upgrades, network infrastructure design flexibility, and clinician and patient information mobility. WAPs enable easy communication among facility guests, patients, and clinicians. WAPs also introduce complexity and uncertainty when wireless devices are allowed access to network connections without preregistration.

Given organizational budgetary constraints and the primary focus on building medical staff capacity, rather than IT staff, segmenting network traffic can reduce the risk of message and signal integrity, in a MITM or identity spoofing attack. By isolating guest and other occasional (i.e., not preregistered) devices to their own, separate network, opportunity for compromise of the facility's protected assets is reduced. Different network segments need firewall separation with activity traceable to individual devices and incidents. The US DHS recommends whitelisting processes, machines, individuals, and data packages that are permissible. Devices that receive, transmit, and/or store patient information should connect through a hardened network connection. Robust access control policies that are well enforced help ensure that machines, individuals, or processes are not allowed privileged access without explicit challenge/response vetting.

The consequences of integrity compromise of wireless medical devices vary according to the specific use case scenario, as indicated in the following examples:

- Patient-focused active medical devices like insulin pumps can be instructed to deny, modify, or deliver treatment. Treatment may include substance administration (e.g., medication, nutrition, oxygen), mechanical intervention (e.g., automatic defibrillation, life support). Loss of integrity in device programming can result in medication delivery errors (too much, too little, wrong medication).
- Patient-focused passive medical devices vary from orthopedic implants to monitoring instruments (e.g., blood pressure and other vital statistics tracking).

- Clinician-focused active medical devices can be used to perform remote surgery or give the clinician real-time information about the patient's condition. Lack of integrity in the messages or work orders sent can result in patient harm.
- Process-focused active medical devices can be instructed to scan medication barcodes, track patient drug allergies, adjust inventory management systems, control medication and organic material safety (e.g., temperature control to protect and preserve blood samples and banks, tissue samples, and organs for transplant).[25]

San Diego–based Independent Security Evaluators performed white hat exercises to test hypothetical attacks against hospital devices, processes, and EHRs. Carried out under conditions that mimicked actual medical facility operational environments, the exercises highlighted common susceptibilities that allowed the test threat agent to:

- Manipulate a passive medical device from outside the network by performing an authentication bypass attack and causing the patient monitor to transmit false information and act erratically;
- Control medicine administration and other workflow processes from a hospital lobby kiosk (that was not on a segregated network) by taking over barcode scanning equipment and causing patient and treatment mismatch;
- Compromise the EHR system to issue improper treatment work orders by launching a cross-site scripting attack that allowed modification of administrator settings, authorized user changes, and then manipulation of patient records; and
- Identify and compromise medical dispensary devices and inventory control systems.[26]

### Medjacking: Insulin Pump Attack

*Hacker Objective:* Research potential for exploitation of wireless medical devices (both active and passive)

*Hacker Technique:* Research public information resources and reverse engineer devices to achieve C&C over them

*Victim Impact:* Curiosity satisfied, Black Hat Conference paper accepted, and future research directions identified

Following in the fine tradition of researchers acting as guinea pigs, one researcher hacked his own insulin pump and glucose monitor, using his knowledge of wireless technologies and following classic attack sequences (reconnaissance and enumeration, intrusion, reconfiguration) to evaluate the difficulty of "capturing" the devices. His methodology is detailed in the paper he delivered at the 2011 Black Hat Conference.[27] Key steps are outlined below. They offer insight into where information about other types of wireless devices can be found and how that information can be put to use.

1. Review user manual for continuous glucose monitor to obtain device signaling frequency, modulation method, packet length, transmission frequency, FCC ID.
2. Download additional performance specs from the FCC website.
3. Research patient documents for the device under the manufacturer's name.
4. Obtain a RF module that operates within same bands as the target devices (Arduino module, in this case).
5. Configure RF module to manually decode transmissions and identify patterns of target devices.
6. Set the logging on the target insulin pump to HIGH to gather sufficient data to reverse engineer the pump's message formats and command codes.

The researcher exercised good judgment and did not complete the attack. His experiment simply assured him that it was doable. It also verified that such devices are vulnerable to replay attacks (given the absence of time stamping or other protection), transmission spoofing, manipulation of sensor data, and even changing the insulin pump configuration settings. As Radcliffe points out in his discussion of future trends, removing human oversight completely from the glucose monitoring and insulin delivery process by automating it end-to-end could leave the devices—and the user—vulnerable to compromise.

Fearing his potential vulnerability to such an attack, former Vice President Cheney asked that his pacemaker implant's wireless capabilities be disabled to thwart hacking efforts. The US Government Accountability Office (GAO) strongly urged the FDA in 2012 to examine mobile medical device susceptibility to malware, unauthorized access, and DoS.[28] The FDA issued its guidance document—"Medical Devices Data Systems, Medical Image Storage Devices, and Medical Image Communications Devices"—on February 9, 2015, but stated that it would not enforce compliance due to the perceived low risk of an actual exploit being performed.[29]

### Hollywood Hospital Hack Attack

*Hacker Objective:* Financial gain through ransomware infection, denial of access to hospital computing resources, and disruption of business processes

*Hacker Technique:* Suspected spear phishing combined with ransomware kit use

*Victim Impact:* Hospital payment of $17,000 ransom for key to release encrypted computing assets, 2 weeks of reduced productivity (due to manual processes used as a workaround), patient inconvenience (due to unavailability of medical tests that required a functioning network)[30]

Detailed information about the actual attack techniques used is not publicly available as of May 2016 because of ongoing investigations by the FBI and LAPD. FBI Cyber Division Assistant Director James Trainor is quoted as saying, however, that ransomware delivery is becoming more sophisticated and less reliant on an employee's clicking a link. Rather, hackers are seeding legitimate websites with the malware to exploit unpatched vulnerabilities on the end-user side.[31] Since wireless devices are generally subjected to less patching scrutiny, they are more likely to be the weak link in security chains.

In this hack, attackers correctly assessed ability and willingness to pay on the part of the victim hospital. Although no PII or personal health information (PHI) was compromised, the attackers encrypted hospital information assets and insisted on payment of a $17,000 ransom in anonymous bitcoin "script" in exchange for the key to decipher EHR information. The hospital senior management elected to pay the ransom before contacting authorities, but then was also very forthcoming about sharing information about the attack to others. From the attackers' perspective, demanding ransom is less risky and more profitable than is selling stolen identities and other information on the black market or engaging in fraudulent activities. It was a quick hit attack with a good profit, especially given the going price tag for ransomware kits (in the $3000 range) that require minimal expertise and feature multiple options for hackers.

## TAKEAWAYS

- Keep software and security patches up to date.
- Change manufacturer default settings and impose robust password policy.
- Segregate patient-critical, clinician-critical, and procedure-critical communication channels behind firewalls. Do not allow hospital visitor, guest, or vendor communications on the same network segment. Deploy virtual LANs for different devices.
- Enforce least privilege, separation of duty, and role-based access control policies.
- Restrict access to WAPs to whitelisted users.
- Implement white and black lists of executable files.[32]
- Create a buffer zone so that data generated or captured by wireless devices can be abstracted for transmission to clinicians in one dedicated environment, but device access is restricted to another dedicated environment. [33] Device control communications should not travel over the same network segment as device data communications.

- Educate staff regularly about the responsible use of technology and communications.
- Medical facilities can mitigate exposure to disrupted work processes, reduced productivity, and information asset kidnapper demands by becoming more resilient. Deploy a well-designed business continuity plan that includes data backup and recovery to an offsite or out-of-band location.
- Evaluate the decision about whether to pay ransom demand on a case-by-case basis. The FBI has given ambivalent guidance.
- Look to the example of Ottawa Hospital and its preparation to survive such a denial of access attack and its ability to remove ransomware that was installed.

## EHRs, MEDICAL DEVICES, AND ICS

### Availability

Less emphasis has been placed on the automated systems that control power, water, and equipment than on systems that contain patient information or medical devices. These automated systems, both associated with general building maintenance and medical-specific equipment, also have an impact on patient health. They are also significant to facility and organizational health. As was discussed in Chapter 5, WAPs in Commercial and Industrial Contexts, these systems were not initially designed to be interconnected via IP networks. Security mechanisms have not been baked in from the start. Hardening these systems to make them more impervious to inappropriate or unauthorized access thus has to be mindful, with attention paid to all people, processes, and technologies involved. ICS in medical facilities are used in patient-critical, clinician-critical, organization-critical, and facility-critical applications.

Medical environments have a tighter tolerance for performance than many industrial environments that use ICS. Whereas the equipment in the majority of industrial/commercial environments must be tolerant to variations in temperature, air quality, vibration (with the exception of precise manufacturing facilities like clean rooms for chip processors), the surgical, imaging, and diagnostic equipment in medical facilities are generally intolerant to such variations. Environmental conditions outside recommended parameters may compromise equipment reliability and even availability.

Consistent power is essential to support active medical devices (AMDs, which do not have their own internal power supply), surgical and other treatment/diagnostic equipment (e.g., equipment use for anesthesia, electrocautery, ultrasound, X-ray, MRI), EHR use, communications network infrastructure, and building environmental controls. All are critical to successful patient and clinician outcomes. Building environmental controls cover lighting, HVAC, security, and air handling. In a hospital environment, it is essential that air is exchanged regularly to reduce contamination and infection spread.[34] ASHRAE 170 (2008) and CDC guidelines (2005) recommend an air change per hour (ACH) rate of at least 12 ACH. By contrast, a home air conditioner typically runs at 0.5–2 ACH—and a clean room environment recommended ACH can vary between 10 and more than 600 ACH. The ISO outlines three environmental "states" for measuring ACH: as-built (finished but empty space), at-rest (when instruments and equipment are introduced), and operational.[35]

As with ICS discussed in Chapter 5, WAPs in Commercial and Industrial Contexts, many medical devices require 24 × 7 availability. This can complicate upgrade and testing, especially because common medical industry practice is to outsource support for, and management of, devices to the manufacturers' own external technicians. Hospital IT staff do not tend to be device specialists with the necessary training to configure devices. In addition to challenges with respect to the "always on" nature of these devices, hospital staff cannot risk modifying FDA-approved devices, even to implement security controls.[36]

Although the FDA is not enforcing compliance with guidelines for medical device security, some researchers and analysts question the wisdom of that decision. One white hat team examined the compromise potential—and condition—of devices like X-ray equipment, blood gas analyzers (BGAs), and picture archive and communications systems (PACS) in an actual hospital setting. Used as unsecured WAPs inside the hospital's trusted environment, the infected devices opened up backdoors to systems from which hackers could pivot and move laterally through hospital networks. The hospital name was anonymized in the study, but one of the attacks looked at is summarized below.

## PACS Pivot Attack

*Hacker Objective:* Extrude PHI; deliver malicious payload (Zeus and Citadel) to achieve C&C over networked devices

*Hacker Technique:* Hospital insider surfed a malicious website from which redirection was to another malicious link; java exploit loaded into user's browser; attacker remotely commanded a malware injection to open backdoor into network; lateral movement through network to PACS; pivot through unprotected PACS to other network assets; encrypted extrusion of PHI through TCP port 443 (used for secure socket layer or SSL)

*Victim Impact:* Extrusion of confidential records to a location in China; infection of a key nurse's workstation compromise of the PACS, which is a critical resource for patient diagnosis, clinician use both on-site and in off-site offices, and organizational reporting and recordkeeping; persistent movement through the network and attempted connection to an external C&C point

In another hospital setting, this white hat group found Zeus and Citadel malware being used to capture network passwords through compromised BGAs, which enabled backdoors and lateral movement. Infection of an IT workstation was also enabled due to the infection of one of the hospital IT department's workstations. An "upgraded," masked version of the net.sah.worm.win32.kino.kf [sic][37] worm had also propagated itself, but was not identified by the hospital's installed cyber defense tools.[38]

> The medical devices themselves create far broader exposure to the healthcare institutions than standard information technology assets. It is the ideal environment upon which to launch persistent attacks with the end goal of accessing high value data. This exposure is not easily remediated, even when the presence of malware is identified conclusively.[39]

Hospitals are challenging environments with respect to infections, both biological and digital. To date, organizational security investments tend to be focused on maintaining EHR confidentiality: data breaches receive undesirable publicity and undermine organizational reputation in comparison to other competing medical organizations.

Failure to comply with HIPAA requirements about protecting patient records from inappropriate access results in punitive fines that destabilize financial capacity. Even when there is no data breach, as in some ransomware incidents, EHR availability is denied and staff productivity is reduced when manual workarounds are required. Patients also suffer inconvenience when they have to visit more remote facilities for required tests or procedures.

## TAKEAWAYS

- Consider patient well-being when prioritizing security investments.
- Improved data governance, patient identification conventions, and protected record management (perhaps through encryption) can serve the security objective of confidentiality.
- Acknowledge the susceptibility of medical devices to compromise and implement mechanisms to ensure their integrity and availability. Differentiate between organization-critical applications whose failure can jeopardize the well-being of the entire organization; building-critical applications whose failure can jeopardize automated environmental, surveillance, water, power, and access mechanisms; and clinician- and patient-critical applications whose failure can jeopardize patient health and survival.

- Evaluate and compensate for the risks associated with the entry of access credentials in the presence of patients, guests, or other nonstaff members. The security of mobile systems and physician and nurse workstations may be especially at risk.
- Enforce written policies to prohibit use of certain insecure protocols and, where practical, include a whitelist of protocols that are permitted.
- Implement network segmentation by function and user. Monitoring devices should be segregated from operating control devices, and both should be segregated from general purpose computing devices. Public access should be segregated from hospital access.
- Use different (i.e., unique and not shared between staff members) WPA2 passwords to implement some minimal encryption of traffic on each network segment.
- Perform regular audits of security design and compliance, including configuration validation. Use external parties for audits as appropriate.[40]
- Become familiar with the National Healthcare and Public Health Information Sharing and Analysis Center (NH-ISAC) to receive and exchange information about recently discovered or even ongoing attacks.[41]

## CONCLUSION

The volume of patient and sensor information transmitted, the plethora of interconnected wireless devices, and the limited internal staff resources for validating EHRs and managing devices increase the likelihood that a compromise at the WAP level will proceed undetected. Device manufacturers are not, however, compelled by the FDA to comply with the guidelines for enhancing device security that the FDA developed at the behest of the GAO. Information confidentiality still leads regulatory concerns about security objectives. Failing to prevent compromise of patient information assets will result in significant financial penalties. Meanwhile, lack of integrity in medical device signaling and function can cause serious harm to patients themselves. Likewise, medical device availability is essential for maintaining patient-critical, clinician-critical, facility-critical, and organization-critical infrastructure and processes. Legislated security objective priorities are not completely aligned with patient well-being.

## ENDNOTES

1. Capnography is used to monitor $CO_2$ under conditions requiring anesthesia or intensive care. Definition from *Wikipedia*. Retrieved from https://en.wikipedia.org/wiki/Capnography.
2. BCC Research (January 2014), "Wireless Electronic Health Records: Technologies and Global Markets," *BCC Research*. Retrieved from http://www.bccresearch.com/market-research/healthcare/wireless-electronic-health-records-hlc147a.html.

3. CDC (March 2016), "National and State Healthcare-Associated Infections Progress Report," *Center for Disease Control*, p. 6. Retrieved from http://www.cdc.gov/hai/pdfs/progress-report/exec-summary-haipr.pdf.
4. Some hospitals have sought to reduce the amount of disinfectant used by watering down products. Romania's Hexi Pharma has done this for years. This practice is associated with thousands of deaths. Source: VICE Romania (10 May 2016), "The Government Cover-Up Behind Romania's Filthy Hospitals," *Vice Media LLC*. Retrieved from http://www.vice.com/en_ca/read/romania-hospitals-infection-health-minister-resignation-876.
5. Patchen Barss (3 March 2016), "#RSAC: Security of Networked Medical Devices must Accommodate Real-World Medical Practice," *InfoSecurity*. Retrieved from http://www.infosecurity-magazine.com/news/rsac-security-medical-devices/.
6. Bob Zemke and Ali Youssef (26 November 2013), "Commentary: Is Your Network Ready? New Challenges Ahead for Hospitals in Managing Wireless Medical Devices," *Modern Healthcare*. Retrieved from http://www.modernhealthcare.com/article/20131126/NEWS/311269956.
7. Institute for Critical Infrastructure Technology (January 2016), "ICIT Brief: Hacking Healthcare IT," *ICIT*. p. 6. Retrieved from http://icitech.org/wp-content/uploads/2016/01/ICIT-Brief-Hacking-Healthcare-IT-in-20161.pdf.
8. TrapX Labs (7 May 2015), "Anatomy of an Attack," *TrapX Security Inc.* Retrieved from http://deceive.trapx.com/rs/929-JEW-675/images/AOA_Report_TrapX_AnatomyOfAttack-MEDJACK.pdf?alild=1116933.
9. Lucas Mearian (8 December 2015), "Virtual Care to Become the Norm," *Computerworld*. Retrieved from http://www.computerworld.com/article/3013013/healthcare-it/cyberattacks-will-compromise-1-in-3-healthcare-records-next-year.html.
10. Others include the Children's Online Privacy Protection Act of 1998 (COPPA) and the Financial Services Modernization Act of 1999, known more familiarly as the Gramm-Leach-Bliley Act (GLBA). These laws were preceded by the Privacy Act of 1974, the Federal Educational Rights and Privacy Act of 1974 (FERPA), and the Electronic Communications Privacy Act of 1986 (ECPA).
11. Supreme Court Justice Louis Brandeis, *Olmstead v. U.S.*, 277 U.S. 438 (1928). Retrieved from http://www.ala.org/Template.cfm?Section=ifissues&Template=/ContentManagement/ContentDisplay.cfm&ContentID=25304.
12. Howard Anderson (23 February 2011), "HIPAA Privacy Fine: $4.3 Million," *InfoRisk Today*. Retrieved from http://www.inforisktoday.com/hipaa-privacy-fine-43-million-a-3375.
13. US Health and Human Services (16 July 2008), "Resolution Agreement: HHS, Providence Health & Services Agree on Corrective Action Plan to Protect Health Information," *HHS*. Retrieved from http://www.hhs.gov/hipaa/for-professionals/compliance-enforcement/examples/providence-health/index.html.
14. Hunton & Williams (20 February 2009), "CVS Pays $2.25 Million in Record HIPAA Settlement," *Privacy & Information Security Law Blog*. Retrieved from https://www.huntonprivacyblog.com/2009/02/20/cvs-pays-2-25-million-in-record-hipaa-settlement/.
15. Results from Privacy Rights Clearinghouse database query performed 23 May 2016. Retrieved from https://www.privacyrights.org/data-breach.
16. Cynthia Burghard, Massimiliano Claps, Lynne Dunbrack, Nino Giguashvili, Judy Hanover, Sven Lohse, Alan S. Louie, Ph.D., Scott Lundstrom, Ashwin Moduga, Eric Newmark, Jeff Rivkin, Silvia Piai, Michael Townsend, Leon Xiao (November 2015), "IDC FutureScape: Worldwide Healthcare 2016 Predictions," *IDC*. Retrieved from https://www.idc.com/research/viewtoc.jsp?containerId=259908.
17. A watering hole attack is, in effect, a form of targeted marketing or a malicious honeypot. The attacker determines likely interests shared by the victim group being targeted, then hacks a website deemed attractive to that group and injects malicious software. Some successful watering hole attacks compromise legitimate websites, for example, the iOS developer forum used by mobile developers for Facebook, Apple, and Twitter. See Michael Mimoso (20 March 2013), "Why Watering Hole Attacks Work," *Kaspersky Lab Threatpost*. Retrieved from https://threatpost.com/

why-watering-hole-attacks-work-032013/77647/. The FBI has used a watering hole attack strategy in its campaign against child pornography. Athletic event websites are also used for watering hole attacks, as was the hookup website Ashley Madison.

18. Dan Goodin (28 July 2015), "Group that Hacked Anthem Shared Weaponized 0-Days with Rival Attackers," *ArsTechnica*. Retrieved from http://arstechnica.com/security/2015/07/group-that-hacked-anthem-shared-weaponized-0-days-with-rival-attackers/.
19. Ponemon Institute (5 May 2014), "Ponemon Institute Releases 2014 Cost of Data Breach: Global Analysis," *Ponemon Institute Press Release*. Retrieved from http://www.ponemon.org/blog/ponemon-institute-releases-2014-cost-of-data-breach-global-analysis.
20. Dan Munro (31 December 2015), "Data Breaches in Healthcare Totaled Over 112 Million Records in 2015," *Forbes*. Retrieved from http://www.forbes.com/sites/danmunro/2015/12/31/data-breaches-in-healthcare-total-over-112-million-records-in-2015/#b724ece7fd5a.
21. Michael Hiltzik (6 March 2015), "Anthem is Warning Consumers about its Huge Data Breach. Here's a Translation," *Los Angeles Times*. Retrieved from http://www.latimes.com/business/la-fi-mh-anthem-is-warning-consumers-20150306-column.html.
22. Mary A. Chaput (24 March 2015), "Calculating the Colossal Cost of a Data Breach," *CFO*. Retrieved from http://ww2.cfo.com/data-security/2015/03/calculating-colossal-cost-data-breach/.
23. Steve Ragan (30 June 2015), "Memo Reveals 312 Different Hashes for the Sakula Malware," *CSO*. Retrieved from http://www.csoonline.com/article/2942601/disaster-recovery/fbi-alert-discloses-malware-tied-to-the-opm-and-anthem-attacks.html.
24. AHiMA, "Integrity of the Healthcare Record: Best Practices for EHR Documentation (2013 Update)," *American Health Information Management Association*. Retrieved from http://library.ahima.org/doc?oid=300257#.V0d5C2OMBsM. This document includes multiple examples of bad practices.
25. Independent Security Evaluators (23 February 2016), "Securing Hospitals: A Research Study and Blueprint," *ISE*. Retrieved from https://www.securityevaluators.com/hospitalhack/.
26. Independent Security Evaluators, op. cit.
27. Jerome Radcliffe (2011), "Hacking Medical Devices for Fun and Insulin: Breaking the Human SCADA System," *Black Hat Paper Submission*. Retrieved from https://media.blackhat.com/bh-us-11/Radcliffe/BH_US_11_Radcliffe_Hacking_Medical_Devices_WP.pdf.
28. Lisa Vaas (22 October 2013), "Doctors Disabled Wireless in Dick Cheney's Pacemaker to Thwart Hacking," *Naked Security by Sopohos*. Retrieved from https://nakedsecurity.sophos.com/2013/10/22/doctors-disabled-wireless-in-dick-cheneys-pacemaker-to-thwart-hacking/.
29. Michael Bassett (24 February 2015), "FDA Calls it Quits on Regulating Medical Device Data Systems," *Radiology Business*. Retrieved from http://www.radiologybusiness.com/topics/policy/fda-will-no-longer-enforce-regs-medical-device-data-systems.
30. Brian Barrett (16 February 2016), "Hack Brief: Hackers Are Holding an LA Hospital's Computers Hostage," *Wired*. Retrieved from https://www.wired.com/2016/02/hack-brief-hackers-are-holding-an-la-hospitals-computers-hostage/.
31. Federal Bureau of Investigation (29 April 2016), "Incidents of Ransomware on the Rise," *FBI Website*. Retrieved from https://www.fbi.gov/news/stories/2016/april/incidents-of-ransomware-on-the-rise/incidents-of-ransomware-on-the-rise.
32. Kacy Zurkus (6 April 2016), "Hospitals Hacks Put Patient Health at Risk," *CSO Online*. Retrieved from http://www.csoonline.com/article/3051094/security/hospitals-hacks-put-patient-health-at-risk.html.
33. Zurkus, op. cit.
34. Farhad Memarzadeh and Weiran Xu (2012), "Role of Air Changes per Hour (ACH) in Possible Transmission of Airborne Infections," *BUILD SIMUL (2012)* Vol. 5, pp. 15–28. DOI:10.1007/s12273-011-0053-4. Retrieved from http://orf.od.nih.gov/PoliciesAndGuidelines/Bioenvironmental/Documents/RoleofACHinTransmissionofAirborneInfections508.pdf.
35. Cleanroom Information: Terra Universal Website. Retrieved from https://www.terrauniversal.com/cleanrooms/iso-classification-cleanroom-standards.php.

36. TrapX, op. cit.
37. Based on my research, I believe the correct spelling is net.sah.worm.win32.kido.kf, a Conficker strain.
38. TrapX, op. cit., pp. 13–14.
39. TrapX, op. cit., p. 15.
40. Independent Security Evaluators, op. cit.
41. Kelly Jackson Higgins (12 February 2015), "How Anthem Shared Key Markers of its Cyberattack," *Information Week Dark Reading*. Retrieved from http://www.darkreading.com/analytics/threat-intelligence/how-anthem-shared-key-markers-of-its-cyberattack/d/d-id/1319083.

CHAPTER 7

# Hacking Wireless Access Points: Governmental Context

The widespread anxiety about Big Brother government is perhaps less related to remembered childhood sibling rivalries and Orwellian dystopian visions than it is to our inability to escape interactions with government. This may be why we experience less collective outrage when a retailer is breached and our personal information is exposed than when a government agency experiences a similar incident—and consequences. We can choose to patronize another retailer. We cannot choose to patronize another government (unless we relocate or emigrate). Unfortunately, the number of US breaches reported publicly for all levels of government between 2005 and mid-2016 was 314, with at least 148,491,010 records (not individuals) compromised, according to the Privacy Rights Clearinghouse.[1]

And yet as citizen consumers we increasingly insist on digital options for filing government reports, receiving government benefits, and retrieving government-curated information and services. Convenience, as discussed throughout these chapters, often relies on the use of WAPs and devices, in spite of their acknowledged vulnerabilities. In making our individual risk calculations, we often assign less weight to safety than to speed. The traditional cost–time–quality/scope triangle of competing factors for producing or achieving outcomes is thus skewed toward time at the expense of quality, which in data terms is associated with the security objectives of CIA.

Public/private partnerships are touted as a logical way of moving forward, especially given interdependent interests. Those partnerships will include state and local government agencies increasingly on the public side. Organizations that work with federal government clients, in particular, should be keenly aware of the Federal Information Systems Management Act of 2002 (FISMA) and the guidelines developed by the US National Institute of Standards and Technology (NIST).[2] The suite of guidelines issued by NIST offer useful tips to organizations (or individuals, really) who have information assets that they value. The National Checklist Program provides a repository for configuration guidance for specific IT products. The process for contributing a checklist is described in NIST SP 800-70 Revision 3.[3]



Hacking Wireless Access Points. DOI: http://dx.doi.org/10.1016/B978-0-12-805315-7.00007-3

The NIST team envisions checklists for component and device configuration that are organized along different environmental dimensions: standalone (i.e., individually managed), enterprise or managed (centrally controlled), custom (e.g., specialized security-limited functionality), legacy (also a custom environment, but one in which system components might not be adaptable to, or interruptible for, upgrade to a more secure configuration), and security-specific. The checklists are intended to be fine-tuned for the local environment after a methodical identification of functional needs, threats and vulnerabilities, and security needs.

Private sector companies that fail on security dimensions when working with government entities suffer damage to their reputations—and to their very financial viability, as seen in the aftermath of the 2014–2015 breaches at the US OPM and the May 2013 compromise of 10,000 US DHS employee records. US Investigations Services (USIS) lost both OPM and DHS government contracts, which comprised about one-third of parent company Alegrity's revenue. The DHS breach was traced to a vulnerability in the system it used for processing background investigations for security clearances. USIS was already under scrutiny from a whistleblower's reporting of deficient background investigation processes (and even "flushing" and "dumping"). USIS is the company that vetted ex-NSA employee Edward Snowden and also Aaron Alexis (the solo gunman in the 2013 Washington Navy Yard shooting spree that killed 12 people).[4]

KeyPoint Government Solutions was tapped by OPM to step in after USIS' contract was terminated. Throughout the June 2015 Congressional hearings with the now-former technology leads at OPM, members of the House Committee on Oversight and Government Reform expressed frustration at the failure of OPM Director Katherine Archuleta to respond directly to questions. This frustration was expressed vividly by Congressman Lynch who said:

> You know what, this is one of these hearings where I think I am going to know less coming out of this hearing than I did when I walked in because of the obfuscation and the dancing around that we are all doing here … I wish that you were as strenuous and hard working at keeping information out of the hands of hackers as you are [at] keeping information out of the hands of Congress and Federal employees. It is ironic. You are doing a great job stonewalling us, but hackers not so much.[5]

For defense-in-depth to work, vulnerabilities throughout the OSI stack have to be mitigated and safety mechanisms built in that reflect the entirely reasonable assumption that an attack somewhere will occur. Similarly, vulnerabilities throughout the supply chain or business partner network must be mitigated and safety mechanisms built in that segregate functionality, coordinate detection, and collaborate on prevention. If each component of the system is appropriately robust down to the data level, the resiliency of the whole is promoted. The OPM failure offers an example of interconnected failures in terms of:

- Process and Policy
  - No formal incident response procedure
  - No regular auditing of configuration settings for workstations, servers, and databases
  - No formal process for auditing physical access privileges
  - No requirement for personal identity verification (PIV) cards for multifactor authentication to access key IT systems
- Technology
  - No standard configuration of firewalls
  - No outbound web proxy[6]
- People
  - Limited in-house staff for security work (contractors manage 22 of 47 major IT systems at OPM, which were not monitored by OPM's SIEM tools as of October 2014)[7]
  - Ineffective leadership, planning, and organizational skills

For hackers, pointillist attention to these background services is desirable: Confusion at jurisdictional boundaries offers attack opportunities within any organization. The successful exploit against the US OPM is an example of this. Year after year, Inspector General (IG) reports identified problems in the decentralized deployment and management of protected data security mechanisms. Gaps in security proliferated, and the agency network was successfully compromised in two distinct incidents that resulted in the exposure of at least 21.1 million records of current and past government employees and job applicants, including those requesting top secret security clearance. In addition to text-based information, some 5.5 million fingerprints were also extruded by hackers, presumably based in China.

## ATTACK CHRONOLOGY

### Doxing Attack for PII Extrusion: *OPM, Department of Interior (DoI), DHS*

*Hacker Objective*: Obtain deep background information and biometric details about government officials (including those with security clearances)

*Hacker Technique:* Multilayer, multiple techniques (reconnaissance through supply chain vendor; exploitation of multiple unpatched software vulnerabilities, e.g., Java, Windows XP, COBOL; lateral movement through network; malware insertion; data extrusion)

*Victim Impact (Individuals):* Foreign state access to PII and biometric details of more than 21 million prospective, current, and past government officials, including those engaged in national security activities; release of biometric information (5.6 million fingerprints) that could be used to fool multifactor authentication systems or implicate individuals in untoward activities; unknown potential for future impact (e.g., identity theft, impaired personal and family safety, compromise of online accounts)

*Victim Impact (OPM Director and OPM CIO):* Both individuals suffered serious, public damage to their professional reputations. The OPM Director resigned from service and the OPM CIO retired from service.

*Victim Impact (Indirect; Contractor Organizations)*: Both directly involved contractor organizations suffered financial loss. USIS lost business-sustaining contracts with the US Government. Contractor KeyPoint's professional credibility was damaged.

*Victim Impact (US Government):* Exposure of agent network to potential compromise and loss of trust with respect to current and future employees

The OPM data breach does not so much illustrate the possibility of cascading failures as illuminate a cataract of existing failures. The preconditions for the successful attack are legion; it was a matter of time. And yet, some of the basic flaws are common to other agencies, so this type of incident is not likely the last.

Initial reconnaissance appears to have occurred through a hack of a contractor's system to obtain privileged credentials. Inadequate security protection of the OPM system had been documented by the OPM's IG for years in its annual FISMA audit reports. Since 2009, the IG had identified problems with OPM's information security management structure as "material weakness."[8] One analysis has suggested that the publicly reported weaknesses may have encouraged hackers to initiate reconnaissance against systems identified as having inadequate authentication and authorization mechanisms in place.[9]

The collateral damage possible from this attack could include blackmail or pressure on government officials with security clearances due to the highly personal and detailed nature of the records extruded; the damage potential goes beyond identity theft and public embarrassment. This incident highlights the dependence of government agencies on outside contractors—and lack of enforced compliance with FISMA requirements for these contractors. It also reveals the amount of sharing among government agencies with respect to infrastructure technology. The DoI is the repository for the OPM personnel records database through DoI's Interior Business Center, which provides cloud government services for multiple agencies as part of a cost-streamlining effort. Some 150 different agencies use DoI IT services. OPM maintains electronic personnel records for millions of current and past government employees, including Congressional staffers. The electronic version of this system (eOPF) is accessible from OPM (and some other departments) through Internet portals—and, apparently, a gateway used by other agency web servers.[10]

*Summary of Attack and Incident Detection Chronology*[11]

| | |
|---|---|
| May 7, 2014: | Access to OPM LAN (malware installed) |
| July 3, 2014: | Backdoor exfiltration started |
| August 22, 2014: | USIS hacked (DHS, customs personnel records compromised) |
| October 2014: | Pivot to DoI (OPM personnel records database) |
| December 15, 2014: | Data (4.2M records) siphoned |
| December 2014: | Decryption tool implemented |
| April 15, 2015: | Anomalous SSL traffic observed beginning in December |
| April 2015: | DHS CERT notified |
| April 17, 2015: | Loaded SSL traffic data into Einstein (DoI IDS) |
| April 23, 2015: | Observed historical netflow of data |
| April 2015: | Notified Congress |
| June 2015: | Congressional hearings |
| July 9, 2015: | OPM admits to breach of 21.5M SSNs |
| September 23, 2015: | OPM admits to breach of 5.6M fingerprints |

The supply chain weaknesses within the federal government operate whether supply chain members are external contractors or other government agencies. The obsolescence of legacy federal IT infrastructure is significant, as is persistent vagueness about response procedures when indicators of attack (IoA) or indicators of compromise (IoC) are suspected or even actually identified. A 2016 US GAO report found that US DoD policy was unclear about roles and responsibilities and command responsibilities for supporting civil authorities during a cyber incident.[12] This lack of clarity extends to definitions of who is responsible to protect privately owned telecommunications and power infrastructure. According to Norman C. Bay, Chairman of the Federal Energy Regulatory Commission, "If I had a cyber threat that was revealed to me in a letter tomorrow, there is little I could do the next day to ensure that that threat was mitigated effectively by the utilities that were targeted."[13]

In another critical infrastructure area that has received a lot of attention, especially at the state and local (smart city) levels, the public/private partnership for transportation needs work. Vehicle cyber security is an area that has received insufficient guidance from the US Department of Transportation (DoT). "Modern vehicles contain multiple interfaces—connections between the vehicle and external networks—that leave vehicle systems, including safety-critical systems, such as braking and steering, vulnerable to cyberattacks. Researchers have shown that these interfaces—if not properly secured—can be exploited through direct, physical access to a vehicle, as well as remotely through short-range and long-range wireless channels."[14]

Another GAO report found that the majority of federal government IT spending (75%) is dedicated to operations and maintenance rather than on development, modernization, and enhancement, for which spending has declined by $7.3 billion since 2010. Remarkably, some systems (including those concerned with nuclear force operations) still depend on 8-inch floppy disks. It is highly likely that contractors connected with such systems are likewise still tied to outdated (50-year-old) technology. Systems critical for US citizens include the US Department of Treasury individual master files—the authoritative data source for individual taxpayers. It is about 56 years old and no plans for its replacement exist.[15]

Ideally, a cost/risk/benefit analysis will help government decision makers determine when to work on the substructure that supports—both financially and informationally—the complex system of benefits and payments on which our federal government operates. Attacks against the IRS E-filing PIN system, for example, reveal how hackers and other criminals can monetize stolen PII. In the 2016 attack, unauthorized, bot-based attempts were made to generate E-filing PINs for 464,000 stolen SSNs. Although only 101,000 such PINs were successfully generated before the automated process was halted,[16] that still represents many citizens who were mulcted of anticipated tax returns, and furthered loss of trust in the US Government.

At the state and local government levels, the number one spending priority, according to a study by the Center for Digital Government, is cyber security. For states, the next set of priorities is shared-services cloud computing, mobile apps, and IT personnel recruitment. States are incorporating 21st century technology into their infrastructure. Investment in policy, practices, and people should be in alignment with these changes. Digital literacy and augmentation of staff skills still rank lowest in terms of priorities, however.[17]

## Cyber Storm

Many Federal, state, and local government agencies and private sector organizations conduct periodic (e.g., annual) disaster recovery exercises to train staff in what should be done during emergencies and to identify and correct problems encountered in the simulations so that they do not occur during actual emergencies. Analogous to these disaster planning exercises, the DHS has been conducting a series of cyber security exercises, called Cyber Storm, since 2006.[18] Cyber Storm is a Government-led, full scale, cyber security exercise involving international, Federal and State governments, and private sector organizations. The purpose of the exercises is to exercise and evaluate response, coordination, and recovery mechanisms in reaction to simulated cyber events. Exercise scenarios include affecting or disrupting infrastructure within the energy, information technology, transportation, and telecommunications sectors. Findings showed where communications worked effectively and where communications and planning could be improved.[19]

As security becomes a larger issue for all organizations, it is anticipated that exercises like Cyber Storm will become as common as disaster planning exercises. At a minimum, all organizations should review the Cyber Storm After Action Reports to identify areas they may need to consider in their security plans.

## Federal Government Takeaways

The OPM breach has brought renewed attention to securing information in Federal agencies. Laptops employ common access cards (CACs)[20] or PIV[21] cards to create VPNs to agency networks. Government personnel and contractors are required to undergo security briefings/courses on a yearly basis. These courses identify how data can be compromised (e.g., theft of cellphone or tablet, shoulder surfing), countermeasures (e.g., password-protected cellphone, encrypted data on tablet), and what to do when a (possible) breach occurs.

Contractors are required to provide a security plan, typically based on NIST SP800-53, to identify how they are protecting PII in their systems that support the government. In addition, agencies are starting to levy a requirement on contractors that, if their systems are breached, the contractor is responsible for credit monitoring for all potentially impacted individuals for three years. These costs can run into the millions of dollars after a breach.

Separation of systems/information accessible by the Internet from those with a "For Official Use Only" or higher sensitive information classification still appears to be a work in progress. Another area where security work is required relates to spear phishing. When asked on the phone for some information associated with a person, how does the customer service person verify that the request is valid and should be honored?

## State Government

In the late 1990s and early 2000s, technology-based economic development was being widely promoted to add more dimension to the financial resiliency of US states, especially among those states in the economically fatigued regions that were losing jobs and momentum: the rust belt

(AKA manufacturing and steel belts) and the cotton belt. States along both coasts seemed to have the right combination of telecommunications infrastructure, educated workforce, innovative companies, risk capital, and geographic appeal to compete into the 21st century. In particular, the latter states appeared less exposed to competition from labor-rich and less-regulated economies (especially the BRIC countries: Brazil, Russia, India, China) with respect to historical economic bastions like manufacturing, food processing, and customer service centers. Frequently heard around economic development roundtables were buzzwords like *ubiquitous computing, always on, 24/7, online/inline, collect once/use many.* State governments were asked to share citizen information across organizational boundaries, which sounded easy until innovators realized the consequences of little or no data governance. Naming and spelling conventions and identification fields were not standardized, allowing multiple entries for the same person or address that appeared, to the system, as unique. Existing policies for sharing information did not exist or were prohibited. Smartphones were not yet available—but people happily carried their five-pound (at a minimum) laptops to seek out Wi-Fi hotspot, which were mapped and configured for convenience.

Some incumbent telecommunications companies tried to obstruct the implementation of free public wireless networks, even as the cost of wireless base stations dropped from about $1000 in 1999 to $100 in 2003.[22] Warchalking, which is essentially marking locations with WAPs, emerged as a grassroots effort in 2002 to share signals in the early days of the sharing economy. The pavement near buildings where signals could be picked up was marked with symbols reminiscent of Depression-era hobo signs, and the warchalking practice itself likened to trainspotting or planespotting (hiking trail cairns come to mind as well). Controversy arose over whether subscribers to wireless services, including cities and retail merchants, were allowed to invite others to share signal for free. Service providers like AT&T Broadband and Time Warner Cable sent letters threatening account termination to wireless subscribers who claimed no harm/no foul if others within 300 feet of their access point could pick up signals.[23]

States competed to define—and deliver—eGovernment services. The annual Digital States Survey ranked states according to their relative sophistication with respect to delivering citizen to government (C2G) services, even if these services were implemented on legacy platforms that were largely dependent on mainframes that had already exceeded their end of life expectancy. Without adequate budgetary resources for technology refreshment so that the core foundation was structurally sound, as it were, convenient applications for citizens and governments alike (government-to-government or G2G) were implemented. Not surprisingly, internal processes, policies, and employee training were misaligned with new attack surfaces created by the explosion in eGovernment services. Systems were compromised. Data breaches occurred—and state-based IT teams began to focus more on protecting what was already in place rather than rolling out new initiatives. Within this technological environment, the National Association of State Chief Information Officers (NASCIO) released a comprehensive report[24] about the risks of wireless use with recommendations that are still, and perhaps more, relevant today:

- Address wireless practices specifically in policy documents and enforce effectively them at all levels of the organization.
- Specify a WAP policy that includes regular status checks and checkpoints between each WAP and the Internet: firewalls, IDS, VPNs, authentication requirements. Authentication should be traceable individually, possibly multifactor (depending on the complexity of the network and how deeply segmented it is), and validated with respect to system file activity on a regular basis to identify anomalies and precursor behavior.
- Configure WAPs securely by changing default SSID names (information about default names is available online along with specifications) to a name that does not reveal excessive information about WAP functional role within the network, use robust encryption (for 2016, at least WPA2), disable unnecessary services (e.g., Telnet, SNMP).

States and private industry largely support the infrastructure framework within which cities implement their smart city initiatives. With their larger responsibilities for public safety and utilities, a recent NASCIO survey indicates that only about 25% of respondents indicated that discussions about the IoT had begun.[25] State-level oversight of road and highway construction, however, provides an opportunity for coordinating activities across multiple infrastructure and operational elements. For example, states can mandate that highway-widening projects are leveraged as opportunities for deploying appropriate conduits for future or current needs like telecommunications lines, sensors for water movement and capacity, or GIS mapping devices. Rather than funding multiple, sequential digs in highway rights of way to deploy conduits, coordination at the state level can reduce traffic disruption and cost. States can also enable smart city initiatives by consolidating vendor accounts to take advantage of volume discounts for the kind of services generated that are less glamorous but nonetheless essential for maintaining infrastructure health and resiliency. Although such services may run in the background—for example, data backup, recovery, and storage; information security and incident detection; power redundancy; data analysis and reporting; network configuration mapping; asset administration; best practice standards and vendor selection criteria—they create a stronger mesh across the whole interconnected system, streamline support processes and allocation of technical talent, and encourage cost and performance efficiencies.

## Local Government

The US Government has encouraged local innovation through funding programs for infrastructure buildout, such as the various rural broadband initiatives delivered through the US Department of Agriculture (USDA). Between 2009 and 2015, for example, the USDA awarded Community Connect grants worth more than $77 million for broadband buildout in rural areas.[26] Other federal agencies have contributed to infrastructure capacity as well. The US Department of Labor has funded projects to connect workforce components: local labor outlets (e.g., Workforce One and other unemployment organizations), private sector industries, and educational institutions. In particular, funding has gone to community colleges that have the flexibility to design focused curricula that meets industry needs for specifically trained and certifiable technicians, mechanics, and skilled trades. The HHS has encouraged the adoption of innovation by mandating implementation of EMRs and EHRs, as discussed in Chapter 6, WAPs in Medical Environments.

Implementation of continuing refinements of recording guidelines, such as the coding of specific treatments according to the International Classification of Diseases (ICD), are mandated by October 1, 2015.[27] HHS innovations in electronic recordkeeping followed the initiatives launched by the Veterans Affairs Administration and the branches of (active) military services. Interest in this on the part of the latter, the US Department of the Army, for example, received emotional and clinical support when combat duty conditions required accessible medical records that were not tied to a specific geographic location. Paper files were inadequate for those requiring immediate medical attention in combat situations, particularly when previous medical care information was stored in multiple repositories (e.g., hospitals, private practitioners, and medical clinics in diverse locations), and there were no existing communications or information-sharing channels. Such mandates may drive increased demand for infrastructure improvements at the local level.

Seven cities accepted the US Government's "Smart Cities" challenge in the hope of receiving the $40 million award (with a $10 million sweetener from Vulcan, Inc., in addition to other private sector corporate donations) and succeeded through review rounds. These cities focused on citizen mobility issues with an emphasis on making smart apps so that drivers and pedestrians don't have to be inconvenienced. Only one of the seven cities (out of 78 applications) that have passed through the final round seems to be addressing disconnects between commercial traffic (distribution, logistics, shipping—totaling $1.4 trillion in 2014 or 8.3% of annual GDP)[28] and noncommercial traffic. This city's proposed initiative would deploy sensors on traffic lights to improve throughput for freight and other commercial vehicles to alleviate both traffic congestion and the associated pollution released when such vehicles are idling for extended periods of time.[29]

As with the eGovernment initiatives at the state level in the early 2000s, the original exuberance around smart city concepts has calmed down a bit as city officials and citizens recognize the total cost of such initiatives, the untested scalability (and practicability) of some IoT products and services, and the barriers to collaboration that exist between different organizations and communities of interest. Communities are finding incremental opportunities in terms of program reach (e.g., transportation, parking, lighting, water distribution) and geographic coverage (i.e., starting with a specific neighborhood or school campus as the "non-conference room" pilot).[30] Planning and a holistic view are imperative, combined with clear understanding of what information and assets can be valuable to—and used effectively by—bad actors. Robust data governance should be defined before that data are gathered. Who will view that data? How will it be protected and maintained? How will updates to the data be verified and authenticated to ensure that changes are not just symptoms of a spoofing attack? The data governance discussion should be broached—along with concerns about security, privacy, return on investment—and made integral to planning a truly smart city, rather than one that is just sensor-rich: ... "to fully take advantage of IoT, cities must integrate it into existing data strategies while addressing new challenges and continually refining their procedures as they grow these new projects."[31]

Those participating in smart city initiatives have an international resource for information sharing, lessons learned, and recommended standards. The ANSI Network on Smart and Sustainable Cities (ANSSC), launched in 2014, leverages (through monthly webinars and other channels) the collective wisdom and influence of the ISO technical committee 268 (Sustainable Development in Communities), the ITU, IEC, and ISO/IEC JTC 1, in addition to regional and national standards groups in Europe and Asia.[32]

## TAKEAWAYS

Although money is tight, governments can perform a number of low-cost activities to improve their security posture:

- Require that all employees partake in yearly security seminars and/or courses.
- Identify their systems security profile using an approach similar to SP800-53. When done properly, the response to each security objective identified in SP800-53 is not a *yes* or *no* but a description of how the security objective is realized. Review of such a document allows an organization to identify and prioritize weaknesses in their security system.
- Determine what activities should only be performed on "secured" systems, and then institute policies and mechanisms to make this

happen. (Securing a system can be as simple as enforcing the practice that this application only be accessed on specifically designated computers or only over the agency VPN.)

- Rethink what information your customer service representatives can give out and what information may require manager approval.
- Develop a plan (program) to improve security, including goals, costs, and the impact of no action ( the costs, including financial and reputational if the plan is not implemented). Use the results of the SP800-53 analysis in developing your plan.

Security is a process not a destination. As employees become more security aware, the plan can be expanded to include the following:

- Define roles and responsibilities clearly, both inside and outside organizational boundaries.
- At a minimum, plan and execute cross-departmental tabletop exercises of different attack scenarios.
- Map process and information flow as actually practiced today.
- Segment networks to enforce separation of duty/privilege, implement "safety valves," and enhance accountability (activity tracking).
- Include security requirements as part of third-party vendor contracts (whether those vendors are in the private or public sector).
- Assume that some attacks will succeed; encrypt protected data assets that are accessible over networks. Store archival documents on devices that are not accessible through web server gateways.
- Enforce recommended security best practices throughout the supply chain through expansion of selection criteria and inclusion of relevant contractual provisions. Differentiate between "checkbox" compliance and functional compliance when evaluating vendor suitability. Incorporate this into Government Services Administration (GSA) and other federal procurement services.
- Encourage state and local governments to adopt security best practices among vendor selection criteria.
- Leverage public/private partnerships to improve resiliency of WAPs like wireless routers (whether for individual users or multiple users) through improved design and manufacturing, communicate threat information clearly and without reprisal, and participate in interactive attack exercises.

## SUMMARY

From a hacker's perspective, pointillist attention to government and background services is desirable: Confusion at jurisdictional boundaries offers attack opportunities within any environment or administrative ecosystem.

The successful exploit against the US OPM is just one example of this. Year after year, IG reports identified problems in the decentralized deployment and management of protected data security mechanisms. Gaps in security proliferated, and the agency network was successfully compromised in two distinct incidents that resulted in the exposure of at least 21.1 million records of current and past government employees and job applicants, including those requesting top secret security clearance. In addition to text-based information, some 5.5 million fingerprints were also extruded by hackers, presumably based in China.

This pattern in the government sector is similar to what has been witnessed in the private sector. Sony Pictures assumed its systems were secure and protected from undetected infiltration by an external attacker. That was before North Korean President Kim Jong Un took offense at Sony's distribution of *The Interview*, a film satirizing his late father, and released a digital army of hackers to compromise Sony's network, entertainment assets, and senior management's reputation.[33]

At what point should criminal or vengeful behavior be characterized as a threat to governmental fulfillment of its duties to establish justice, insure domestic tranquility, and provide for the common defense? In the following chapter we will discuss issues around WAP functionality and protection in mission environments involving law enforcement, first responder and emergency management, and military activities.

## ENDNOTES


1. Result from a query run June 19, 2016, against the Privacy Rights Clearinghouse database < https://www.privacyrights.org/data-breach/new > using "GOV[ERNMENT]" as the organizational filter and "HACK," "PORT[ABLE]", and "STAT[IONARY]" as the breach type filter. A graphic of significant data breaches reported worldwide since 2005 is available at http://www.informationisbeautiful.net/visualizations/worlds-biggest-data-breaches-hacks/.
2. Kelley Dempsey, Greg Witte, and Doug Rike (19 February 2014), "Security and Privacy Controls for Federal Information Systems and Organizations," *NIST SP 800-53 Revision 4*. Retrieved from http://csrc.nist.gov/publications/nistpubs/800-53-rev4/sp800-53r4_summary.pdf. Also see the Joint Task Force Transformation Initiative (December 2014), "Assessing Security and Privacy Controls in Federal Information Systems and Organizations," *NIST SP 800-53A Revision 4*. Retrieved from http://nvlpubs.nist.gov/nistpubs/SpecialPublications/NIST.SP.800-53Ar4.pdf.
3. Stephen D. Quinn, Murugiah Souppaya, Melanie Cook, and Karen Scarfone (December 2015), "National Checklist Program for IT products – Guidelines for Checklist Users and Developers," *NIST SP 800-7 Revision 3*. Retrieved from http://nvlpubs.nist.gov/nistpubs/SpecialPublications/NIST.SP.800-70r3.pdf.
4. Peg Brickley (22 June 2015), "U.S. may Hit Altegrity for Damages in USIS Whistleblower Case," *The Wall Street Journal*. Retrieved from http://www.wsj.com/articles/SB10181154852263044738504581063990654796106.
5. Transcript of hearings before the House Committee on Oversight and Government Reform (16 June 2015). Reporter M. Jones. Retrieved from https://oversight.house.gov/wp-content/uploads/2015/06/2015-06-16-FC-OPM-Data-Breach.GO167000.pdf.

6. Sean Lyngaas (18 February 2016), "IG Details OPM Contractor's Security Flaws," *FCW*. Retrieved from https://fcw.com/articles/2016/02/18/opm-oig-keypoint.aspx.
7. Sean Gallagher (21 June 2015), " 'EPIC' Fail—How OPM Hackers Tapped the Mother Lode of Espionage Data," *ArsTechnica*. Retrieved from http://arstechnica.com/security/2015/06/epic-fail-how-opm-hackers-tapped-the-mother-lode-of-espionage-data/.
8. OPM Office of the Inspector General (10 November 2015), "Final Audit Report," *Federal Information Security Modernization Act Audit FY 2015 Report Number 4A-CI-00-15-011*. Retrieved from https://www.opm.gov/our-inspector-general/reports/2015/federal-information-security-modernization-act-audit-fy-2015-final-audit-report-4a-ci-00-15-011.pdf.
9. Sean Gallagher (21 June 2015), " 'EPIC' Fail—How OPM Hackers Tapped the Mother Lode of Espionage Data," *ArsTechnica*. Retrieved from http://arstechnica.com/security/2015/06/epic-fail-how-opm-hackers-tapped-the-mother-lode-of-espionage-data/.
10. Ibid.
11. Sean Lyngaas (21 August 2015), "The OPM Breach Details You Haven't Seen," *FCW*. Retrieved from https://fcw.com/articles/2015/08/21/opm-breach-timeline.aspx.
12. Federal News Radio Staff (5 April 2016), "Tuesday Federal Headlines – April 5, 2016," *Federal News Radio*. Retrieved from http://federalnewsradio.com/federal-headlines/2016/04/tuesday-federal-headlines-april-5-2016/.
13. As quoted in the Truman Security Briefing Book 6th Edition (12 May 2014), *Truman National Security Project*. Retrieved from http://briefingbook.trumanproject.org/wp-content/uploads/2014/08/TSBB-6-Cyber-Chapter.pdf.
14. US Government Accountability Office (March 2016), "Vehicle Cybersecurity: DOT and Industry Have Efforts Under Way, but DOT Needs to Define its Role in Responding to a Real-World Attack," *GAO Report 16-360*. Retrieved from http://www.gao.gov/assets/680/676064.pdf.
15. US Government Accountability Office (25 May 2016), "Federal Agencies Need to Address Aging Legacy Systems GAO-16-696T," *GAO*. Retrieved from http://www.gao.gov/products/GAO-16-696T.
16. Internal Revenue Service Statement (9 February 2016), "IRS Statement on e-Filing PIN," *IRS Website*. Retrieved from https://www.irs.gov/uac/newsroom/irs-statement-on-efiling-pin.
17. Justine Brown (24 March 2016), "IT Spending is on the Rise – but Where Will the Money Go?" *e.republic*. Retrieved from http://www.erepublic.com/blog/IT-spending-is-on-the-rise--but-where-will-the-money-go.html.
18. https://www.dhs.gov/cyber-storm.
19. Cyber Storm Exercise Report, September 12, 2006.
20. "The CAC, a "smart" card about the size of a credit card, is the standard identification for active duty uniformed service personnel, Selected Reserve, DoD civilian employees, and eligible contractor personnel." Retrieved from http://www.cac.mil/common-access-card/.
21. "A PIV card is a United States federal smart card that contains the necessary data for the cardholder to be granted access to federal facilities and information systems and assure appropriate levels of security for all applicable federal applications." Retrieved from http://www.ors.od.nih.gov/ser/dpsac/Training/Pages/video.aspx.
22. Terry Schmidt and Anthony Townsend (May 2003), "Why Wi-Fi Wants to be Free," *Communications of the ACM*, Vol. 46, p. 5. Retrieved from http://wenke.gtisc.gatech.edu/wireless-security/p47-schmidt.pdf.
23. Ben Charny (13 July 2002), "Cable Companies Cracking Down on Wi-Fi," *CNET*. Retrieved from http://www.cnet.com/news/cable-companies-cracking-down-on-wi-fi/.
24. National Association of State CIOs (April 2004), "Wireless in the Workplace: A Guide for Government Enterprises," *NASCIO*. Retrieved from http://www.nascio.org/Publications/ArtMID/485/ArticleID/256/Wireless-in-the-Workplace-A-Guide-for-Government-Enterprises.
25. Ben Miller (27 May 2016), "What Role do States Play in the Push to Get Smarter?" *Government Technology*. Retrieved from http://www.govtech.com/state/What-Role-do-States-Play-in-the-Push-to-Get-Smarter.html.

26. US Department of Agriculture (20 July 2015), "USDA Announces Funding for Rural Broadband Projects," *USDA News Release.* Retrieved from http://www.usda.gov/wps/portal/usda/usdahome?contentid=2015/07/0212.xml.
27. American Medical Association (n. d.), "AMA and CMS Set ICD-10 Strategy to Help Physicians with Implementation," *AMA.* Retrieved from http://www.ama-assn.org/ama/pub/physician-resources/solutions-managing-your-practice/coding-billing-insurance/hipaahealth-insurance-portability-accountability-act/transaction-code-set-standards/icd10-code-set/cms-announcement.page?.
28. SelectUSA (n. d.), "Logistics and Transportation Spotlight," *Website.* Retrieved from https://www.selectusa.gov/logistics-and-transportation-industry-united-states.
29. Michelle Davidson (16 June 2016), "Smart City Challenge: 7 Proposals for the Future of Transportation," *NetworkWorld.* Retrieved from http://www.networkworld.com/article/3084455/internet-of-things/smart-city-challenge-7-proposals-for-the-future-of-transportation.html.
30. Adam Stone (2 June 2016), "A New Smart City Model is Emerging," *Government Technology.* Retrieved from http://www.govtech.com/data/A-New-Smart-City-Model-Is-Emerging.html.
31. Stephen Goldsmith (8 June 2016), "5 Key Themes to Consider When Implementing Internet of Things Initiatives," *Government Technology.* Retrieved from http://www.govtech.com/opinion/5-Key-Themes-to-Consider-When-Implementing-Internet-of-Things-Initiatives.html.
32. American National Standards Institute (n. d.), "ANSI Network on Smart and Sustainable Cities (ANSSC): Overview," *ANSI.* Retrieved from https://www.ansi.org/standards_activities/standards_boards_panels/anssc/overview.aspx?menuid=3.
33. Mark Seal (4 February 2015), "An Exclusive Look at Sony's Hacking Saga," *Vanity Fair.* Retrieved from http://www.vanityfair.com/hollywood/2015/02/sony-hacking-seth-rogen-evan-goldberg.

CHAPTER 8

# Noncivilian Government Context

Moving from the civilian or administrative dimension of government—with which citizens, government agencies, and organizations tend to have regular or scheduled interactions—also represents a move away from bureaucratic or calendar-driven deadlines for action, which may be important but not urgent. What we are referring to here as the noncivilian agencies—those whose missions are predominantly action-oriented and event-driven—are characterized by a greater sense of immediacy, criticality, and applicability. Although a considerable amount of time is spent in any government agency on administrative, bureaucratic process work, the agencies involved with public safety, emergency management, and national security operate under tighter schedules and, typically, more stressful conditions. Individual lives (not just livelihoods) are frequently endangered, both for agency representatives and citizens. Their missions carry a different sense of urgency: more acute, harder.

The security contexts discussed in this chapter differ from those in other chapters in terms of attacker resources, legal controls, and societal and individual impact. The level of importance—and complexity—increases dramatically in these environments. These are environments that stress-test technology tools, people, and policies/processes.

## NATIONAL SECURITY

Consider, for example, the attacker resources that exist within the national security context. Attackers/hackers range from small, political hacktivist, terrorist, and mercenary cell groups to coordinated nation-state teams of several thousand technical experts. They likely have access to sophisticated tools, privileged information and credentials (some obtained illicitly), money to subsidize efforts for an unlimited period of time, and immunity from government prosecution, maybe even a degree of anonymity or identity protection. They also have physical resources available to them including safe work spaces and, in some cases, military forces. Their cyber work may be part of a broader kinetic effort that includes the use of physical force, as is discussed



Hacking Wireless Access Points. DOI: http://dx.doi.org/10.1016/B978-0-12-805315-7.00008-5

in the attack scenario that outlines incidents involving Estonia, Georgia, and Crimea.

With respect to legal controls, attackers in the national security sphere may actually act with more impunity than those working at the subnational level. Enforceable laws exist at the latter level that act to discourage large-scale, persistent attacks on critical infrastructure. International law, on the other hand, is fuzzy: What is not explicitly prohibited is permitted. Sovereign state boundaries present barriers to effective prosecution and compliance with some common agreement about the inviolability of cyberspace. Cyberspace seems to evade unanimous agreement about what constitutes international or national space, even in derived extensions to the Geneva and Hague conventions and dedicated efforts like the Tallin Manual and the Cybercrime (or Budapest) Convention. Unlike historical battlegrounds, cyberspace is a man-made construct that is not geographically delimited: Human activity defines it. Engagements in cyberspace may be construed as physically nonviolent (although activity may be seen as a precursor to physical violence). As one scholar observes:

> The gap between use of force and armed attack is already a contentious one, creating disputes about what level and kind of violence meets the 'armed attack' threshold. Cyberspace's unique properties dilute the meaning of these terms further: they enable non-violent electronic incursions - such as data theft, or systems sabotage - on a scale so vast that states' core security interests can be threatened, without any of the immediate kinetic damage traditional attacks produce.[1]

Within the broader noncivilian government genus, agencies in the public safety and emergency management categories—which may include private sector partners among operations staff—function in a more decentralized fashion than those agencies that are responsible for collecting, distributing, and managing common goods (e.g., "civilian" agencies). They are also more decentralized than those involved in promoting and defending national security when considered from a US government perspective although, as discussed earlier, they are highly decentralized when considered from a global perspective.

For this chapter, we will look at entities concerned with national security separately from the first two because of the differences in organizational structure, decision making hierarchy, recruitment/retention strategies, operational work (or response) group size, and enrollment and cultural artifacts. The agencies associated with national defense are characterized by more centralized authority, less behavioral diversity across agencies (as determined by policy and articulated in standard operating procedures), and national rather than local budget support.

These differences lead to significant ripple effects observed when one looks at technology deployments and ensuing communications compatibility across groups. The underlying philosophy of information asset security is also affected because of more centralized planning and (at least hypothetically) adherence to, and longer experience with, a consistent digital security framework.

Law enforcement and emergency management organizations tend to plan and act more locally (i.e., less systemically) than do national security organizations within the United States. In particular, mission-critical, on-the-ground actions are performed within the United States or its contiguous tribal nations: jurisdictions that are subject to common laws. Agencies in the national security arena deploy their operations groups outside these boundaries, typically, and may be subject to a different set of formal laws and informal practices. They also interact with others operating under different legal jurisdictions, in which assumptions about rules of engagement can be dramatically varied and even incompatible with those that operate domestically (i.e., within the United States and its territories and tribal nations).

## PUBLIC SAFETY AND EMERGENCY MANAGEMENT

The lines between public safety and emergency management are blurred since an incident can range from as small as a traffic accident to as large as pandemics, major earthquakes, and nuclear war (Fig. 8.1). The incidents affect individuals, organizations, and infrastructure differently and require various resources for assistance. Incidents do not necessarily happen in isolation from one another also. Priorities must be established along with crisis response windows for common understanding about how incidents are to be managed.

As Federal Emergency Management Agency (FEMA) observes so adroitly: "Incidents typically begin and end locally." Logically, then the most effective responses to such incidents—especially their daily management—will be carried out at the level closest to the source, whether at the geographical, organizational, or jurisdictional level. As the scope of the incident increases so does the likely beneficial "involvement of multiple jurisdictions, levels of government, functional agencies, and/or emergency-responder disciplines." Local resources can be tapped out in natural disaster or emergency events. For example, the 2013 "1000-year flood" in Colorado created an island of one community and resulted in emergency airlifts and hundreds of people unaccounted for a week after the rain stopped. The National Guard was called in to help along with state agency personnel; 14 Colorado counties earned emergency disaster designations. Responses to more complex incidents like this "require effective and efficient coordination across the broad spectrum of organizations and activities."[3]

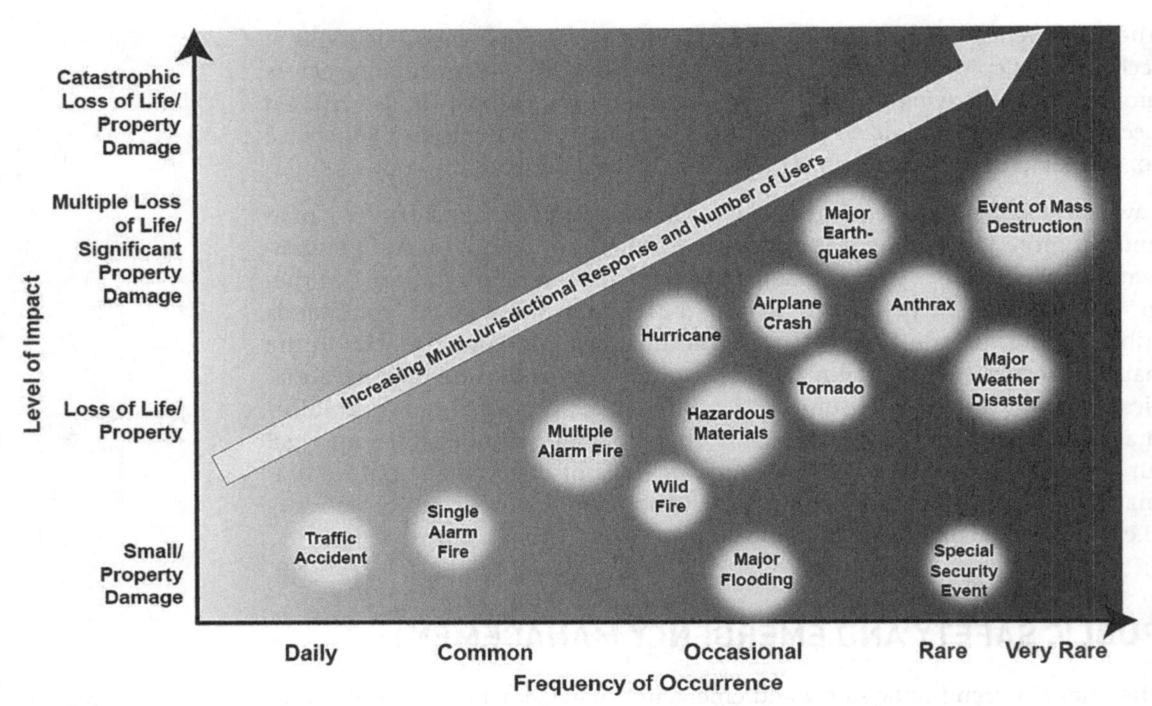


**FIGURE 8.1**
Various events requiring response by National Security and Emergency Preparedness (NS/EP), public safety, or both.[2]

To support effective coordination, the FEMA created the National Incident Management System (NIMS). NIMS provides a flexible but standardized set of incident management practices with an emphasis on common principles, a consistent approach to operational structures and supporting mechanisms, and an integrated approach to resource management to guide departments and agencies at all levels of government, nongovernmental organizations, and the private sector to work together seamlessly to manage incidents involving all threats and hazards—regardless of cause, size, location, or complexity—in order to reduce loss of life, property, and harm to the environment.[4]

## INTEROPERABILITY CHALLENGES: PEOPLE, PROCESS, AND TECHNOLOGY

With decisions about budgets, technology deployments, and specific policies and procedures being made at the substate level for these organizations, coordination across jurisdictions becomes more complicated. In the past, this has led to communications disconnects and technical platform incompatibility, as can be seen by the multiple land mobile radio systems in use across the United States.[5] Procedural and training differences, in addition to communication platform incompatibilities, contribute to the challenges attributed to inadequate interoperability. At some level, interoperability challenges create a de facto DoS situation for first responders. Such a DoS can trigger cascading failures in communication during large-scale emergencies like those witnessed in the hours after 9/11 and during the 2005 Hurricane Katrina operations. For example, the *9/11 Commission Report* records numerous instances of failed communications (in addition to recording that emergency plans developed over the years were inaccessible, locked in a file drawer in a building that was cordoned off near Ground Zero). The US DHS invested billions of dollar in equipment grants for local-level first responders in the first four years after 9/11—between $2.5 billion and $5 billion in FY 2004 alone for digital equipment[6]—but difficult communications during Hurricane Katrina provided clear evidence that the interoperability goals had not yet been achieved.

Under the auspices of the Middle Class Tax Relief and Job Creation Act of 2012, $7 billion in funding was pledged to build out a new First Responder Network Authority (FirstNet). The objective was to override the service provider patchwork of public network telecommunications and develop a single, interoperable broadband architecture to support activities of public safety and emergency management professionals, especially with respect to wireless services. An additional $135 million was included for state and local implementation, in addition to US DoT grant funding to improve local 911 services through the National Highway Transportation and Safety Administration.[7] In January 2016 FirstNet released its request for proposal (RFP) for building the public safety network. The award, which could total $6.5 billion to the winning contractor for this indefinite-quantity-indefinite-delivery business, carries with it access to some 20 MHz of spectrum and requirements for returning investment funds to FirstNet over the course of the 25-year agreement.[8] The RFP includes mobile devices and requirements for hardening the network to guarantee end-to-end data transmission security. Globally, the estimated market for public safety networks is at least 2 billion Euros in 2019.[9]

In the early 2000s, those sitting around state-level emergency planning committee meetings bemoaned the lack of standards across jurisdictions with respect to communications equipment and channels. The US broadband communications infrastructure is much more built out now than in 2005: Connection via 4 G long-term evolution (LTE) is available to 99% of the US population, and 80% of US residents can choose from among at least four LTE providers. Fig. 8.2 shows the build out of US wireless and wired broadband connections between 2009 and 2013.[10]

Of course, great responsibility comes with the communication potential afforded by this broadband growth. A January 2016 letter to FCC Chairman

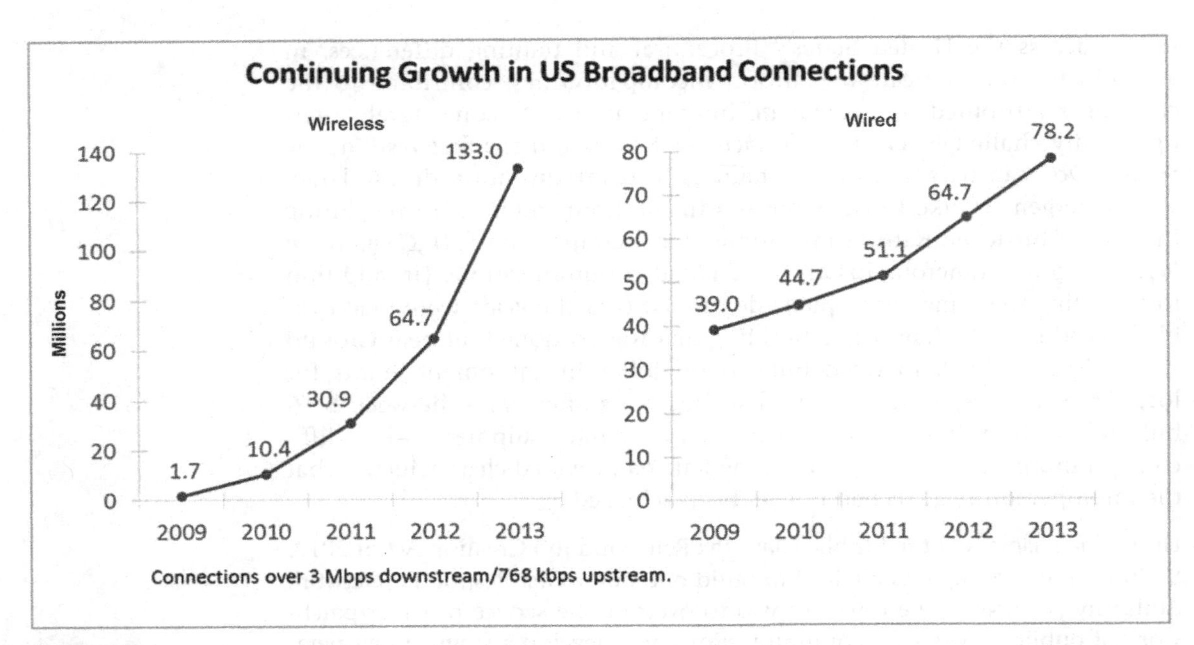


**FIGURE 8.2**
US wireless and wired broadband connections since 2009.[11]

Tom Wheeler has requested that restrictions be placed on ISPs' current "by default" ability to capture and store data on users.[12] White hat researchers and black hat adventurers have verified that ISPs can be hacked. Some of the techniques are accompanied by convenient YouTube tutorials. (This is definitely more sophisticated than signal-jacking a neighbor's ISP connection by leaning against one's apartment door and becoming a human antenna.)

Clearly technology, especially the commercial wireless broadband infrastructure over which much public safety and emergency management passes, has become more sophisticated. Interoperability is not determined just by the network or its components, however: People and policies/procedures are also necessary components of interoperability.

This learning is underscored by observations about what worked and did not work in terms of communication in the immediate aftermath of both 9/11 and Katrina and in summary reports of structured interactive exercises by assembled members of the first responder stakeholder communities (e.g., CyberStorm[13]). The first observation is, in effect, that everything changes in a large-scale emergency. Cascading failures are difficult to predict—although experiencing them does help planners who develop a variety of scenarios. The failures occur in each component of interoperability. In addition to technical failures, failures at the level of policies and procedures and personnel complicate communications. The latter failures have not always received as much attention (at least as evidenced by funding) as have technical failures, in spite of the truism attributed to Sun Microsystem's John Gage, "Technology is easy. People are hard."

## REPRESENTATIVE CHALLENGES: POLICY AND PROCEDURE (AKA PROCESS ENGINEERING)

The joint CyberStorm exercises provide longitudinal information about organizational impediments to effective response in emergency situations. They include the following[14]:

- Complexity causes confusion and requires more coordination horizontally (between public and private sector entities) and vertically (across attack vectors and incidents) (CyberStorm I, p. 5).
- Response procedures tailored for physical crises need to continue to converge and integrate with those developed for cyber events (CyberStorm II, p. 4).
- Increased noncrisis interaction can solidify communication paths and strengthen relationships (CyberStorm II, p. 4).
- Public/private interaction "can be complicated by the lack of timely and meaningful shared situational awareness; uncertainties regarding roles and responsibilities; and legal, customer, and/or security concerns" (CyberStorm III, p. 5).
- Shared situation awareness—a cyber common operating picture (COP)—is critical (CyberStorm III, p. 5).
- It is imperative to catalog and communicate available resources within both the public and private sectors during noncrisis and crisis periods (CyberStorm IV, p. 7).
- Interdependencies, critical systems, and required communications must be documented (CyberStorm IV, p. 6).

## REPRESENTATIVE CHALLENGES: PERSONNEL (AKA HUMAN ENGINEERING)[15]

Certain psycho-physiological responses to emergency situations can interfere with effective communications, thus creating the appearance of an interoperability problem, even though the root cause of that problem is not technical platform incompatibility. Examples observed include the following:

- Reversion to normal usage habits rather than adaptation to emergency context can result in suboptimal use of equipment and networks (e.g., increased communications when communications channel capacity is already taxed).

- "Sensory overload and myopic operational tendencies" are activated when responders go into "automatic mode."
- Cognitive bias can interrupt effective communications when information is incomplete or key elements are inaccurately relayed.
- Stress-induced verbal impediments can complicate voice communications (and cause confusion with respect to the intention and/or identification of the speaker; the FCC recommends text over voice communications in emergency situations because of the former's lower capacity requirement).
- During periods of high-volume, high-stress crisis situations, the user's expectation of and reliance on good communication continues, but the increased pace and load on the radio system, combined with the unique emotional influences present, typically acts to hamper, rather than facilitate, the communications process.
- Personal protective equipment is not designed with access and use of commercial (i.e., general public use) cellular equipment in mind. The reverse is also true: Consumer-grade commercial equipment is not designed with crisis situations, harsh operational environments, and assured communication requirements in mind.
- Idiosyncratic use of codes and other nomenclature can obscure the meaning of messages.

## REPRESENTATIVE CHALLENGES: TECHNOLOGY (AKA NETWORK AND DESIGN ENGINEERING)

Issues surrounding spectrum allocation and network infrastructure ownership are being addressed based on current and near-term need. The prospect of acquiring significant additional spectrum in the 700 MHz band makes FirstNet an even more appealing opportunity for narrowband and broadband providers, especially because that spectrum allocation will not be restricted solely to public safety use by the carrier. Use of wireless broadband is expected to increase, especially given the growth in machine-to-machine (M2M) communications associated with IoT and IoE. Although constituting only 3% of current mobile data communications, M2M is projected to be 20% by 2020 and as much as 35%–47% by 2030.[16]

Predicting future capacity needs is complicated by uncertainty about how various wireless devices will be adopted into use by public safety and emergency management service organizations. In 2016, for example, only about 25% of police officers use body-worn cameras. Broader adoption would expand requirements for tablets (to tag data); upgraded camera technology and procedures to ensure the admissibility of evidence in court (e.g., adequate resolution even under low light and other difficult environmental conditions, chain of custody assurance); security control mechanisms to protect data captured, transmitted, and stored; policy enhancements to preserve privacy, meanwhile respecting public disclosure expectations. Even now, use of smartphones and other mobile devices in the field has launched concerns about how to managing sharing devices among personnel (i.e., having a device "pool" rather than a single dedicated device per individual) and maintaining device application standards.

## Mobile Apps Vetting Process: Test, Approve or Reject[17]

It is important that applications used by public safety and emergency management personnel work as intended during both normal and emergency situations. In addition, it is important that the applications not have "backdoors" that can be used by third parties. To verify the capabilities of both commercially-available and agency-developed applications, the US NIST has developed guidelines for managing mobile apps:

- Define the security requirements for app use.
  - In general, implement controls to "prevent unauthorized functionality" or "protect sensitive data."
  - In specific contexts, for example, implement "location-aware access attempt" (taking advantage of GPS capabilities of apps) or "user-specific functionality" (who can take video or audio recording).
- Use context-appropriate security controls (e.g., a blanket requirement for encrypting data going into the cloud can be overkill if the transmission is via VPN).
- Use a variety of app testing tools plus human interaction.
- Apply "chain of custody" type protection to apps down to the code level in the vetting process (no inadvertent end user license agreement violation!).
- Restrict telephony activities and unauthorized functionality that put PII at risk (e.g., exfiltration to a third party, fake website injection into browsers, pop up banner ads, communication with black-listed or nonwhite-listed sites).
- Disable privacy-compromising functionality (e.g., location and broadcast data, eavesdropping, shared system-level logs).
- Define acceptable network protocol access (e.g., no Bluetooth or NFC).

## Government Emergency Telecommunications Service/Wireless Priority Service

The Government Emergency Telecommunications Service (GETS) and Wireless Priority Service (WPS) are programs administered by the DHS Office of Emergency Communications (OEC). These programs support Executive Order 13618, Assignment of NS/EP Communications Functions:

> The Federal government must have the ability to communicate at all times and under all circumstances to carry out its most critical and time sensitive missions. Survivable, resilient, enduring, and effective communications, both domestic and international are essential to enable the executive branch to communicate within itself and with: the legislative and judicial branches; State, local, territorial, and tribal governments; private sector entities; and the public, allies, and other nations. Such communications must be possible under all circumstances to ensure national security, effectively manage emergencies, and improve national resilience. The views of all levels of government, the private and nonprofit sectors, and the public must inform the development of national security and emergency preparedness (NS/EP) communication policies, programs, and capabilities.[18]

OEC has contracted with key United States commercial telecommunications companies to provide these services to NS/EP users. These telecommunications companies augment their networks with NS/EP-unique functions (such as queueing for telecommunications resources) and specified provisioning of commercial QoS features to provide NS/EP users with a greater probability of call completion when the public networks are congested. Since use of the GETS and WPS features can increase the time for the call to get through the network and be completed, users are told to not to hang up if they "hear" dead air, as the call may still be progressing through the network. GETS and WPS are designed to work under all hazards due to both man-made and natural disasters, including shootings, terrorist attacks, floods, earthquakes, tornadoes, hurricanes, and nuclear war. GETS and WPS are designed to give priority to the call end-to-end, regardless of whether the called party is an NS/EP user or not.

The GETS program is a wireline calling card service that was started in 1995 and obtained its Full Operational Capability in third quarter 2001. As carriers upgrade their networks (e.g., to IP Multimedia Subsystem [IMS] components), the OEC works with the GETS carriers to migrate the GETS features to the new technologies.

The WPS program was started immediately after September 11, 2001, with the goal to provide WPSs to New York City; Washington, DC; and the

2002 Winter Olympics in Salt Lake City. Satellite phones were used for an immediate capability while the government worked with wireless carriers to develop and implement features in nationwide global system for mobile (GSM) and code-division, multiple-access (CDMA) networks. As newer technologies (e.g., universal mobile telecommunications system or UMTS and LTE) were deployed in the WPS carrier networks, the Government worked with the carriers to develop approaches to ensure WPS continuity. In UMTS networks, recognition of a WPS call by the network causes the network to tell the phone to "fallback" to the GSM network. WPS features are currently being developed and deployed in LTE.[19] While this is occurring, users are told how to make their phones fall back to the GSM or CDMA technologies.

Unlike other countries, US carriers cannot preempt public calls for Government emergency use; in addition, the FCC mandates that carriers must make some bandwidth available for public calls during congestion events. For major identified "congestion events" (e.g., Times Square New Year's Eve; Washington, DC, Mall on Presidential Inauguration Day), the OEC coordinates with the wireless providers on their use of deployables (i.e., mobile wireless cellular access points like cell on wheels or cell on light trucks) for on-demand increased network capacity. Deployables may be used in disaster recovery areas where existing infrastructure has been damaged or in places (e.g., wilderness areas) that are typically "off-grid."

Since the WPS features are infrequently used at any component within the carriers' networks, the Government and service providers validate its viability regularly in the following ways:

- Conduct scheduled and ad hoc WPS calls. Government and service provider personnel make WPS calls when they travel, and test the service before and during identified "congestion events."
- Perform periodic configuration audits of all components with WPS features to ensure that operational changes to the carriers' networks do not "turn off" WPS features.
- Perform periodic tests of each component type within a carrier's network to ensure that new features in the component do not impact the functionality of the WPS features.

### *Lesson 1—Network Performance Under Heavy Congestion*

Most carriers design their network to provide acceptable performance under peak hour loads. They recognize that during rare events, such as Mother's Day, their networks may become congested and performance may degrade. It is not cost-effective for them to engineer their networks to support these rare events.

WPS is engineered to work in congestion events in which the offered load to the wireless networks could exceed 20 times the engineered load of the networks. Even though carriers may have vendors stress-test their components to two times the engineered capacity of the device, OEC needs to be assured that the devices will not collapse under greater loads. Stress-testing to 20 times overload is not always feasible, so a mixture of modeling and testing is used. Performance of components under actual stress events is fed back into the modeling and testing process for future use.

### *Lesson 2—Things Change*

It is important to validate the engineering assumptions for a service periodically. When originally designed in 2002, congestion on the random access channel (RACH) was not seen as likely. During the July 29, 2008 Chino Hills earthquake, WPS users claimed that they could not make calls. Analysis by the Government and private industry providers showed a RACH collapse due to "everyone" attempting to make a voice call at the same time. The Government and industry worked on an approach to resolve this problem using automated access class barring (AACB). A RACH collapse again occurred during the Virginia earthquake of August 23, 2011 in most networks. When the Boston Marathon bombing of April 15, 2013 occurred, the AACB feature was being piloted by one carrier; a RACH collapse again occurred on most carriers' networks, causing news reports to say the Government had shut down the wireless networks to prevent additional terrorist attacks.[20] The Government is currently working with private industry on how to migrate this capability into their LTE networks.

### *Lesson 3—Good Intentions Can Lead to Bad Results*

In GSM and CDMA, 911 calls do not have priority over public calls. As currently implemented using LTE, 911 calls and WPS calls have the same priority over public voice calls. This priority allows WPS and 911 calls access to LTE resources before public calls. This can be a very bad thing under 20 times overload.

Providing 911 calls priority over public calls makes sense under normal circumstances, based on the assumption that only one or two 911 calls will be made in a congested cell. In a major event like a magnitude 8 earthquake in San Francisco with extensive damage to buildings and infrastructure, however, the majority of calls are likely to be 911 during the first hours. These calls will get resources to the Public Safety Access Point (PSAP) but then they will get a busy signal or be put on queue.[21] Because there are many more 911 calls during this event than WPS calls, the WPS calls have a much lower likelihood of successfully accessing the congested resources to be completed. And yet, many of the WPS calls support the response to 911 requests. Thus, having

911 and WPS at the same priority provides a poorer response than if WPS had a higher priority to 911.

An access class barring (ACB) mechanism could be implemented to provide priority to WPS over 911. This approach is politically unacceptable to members of the telecommunications industry, however, since they do not want to be sued for being the responsible party for blocking 911 calls (even though the calls will be blocked at the PSAP). Government rulings and regulations from the FCC will be required to solve this issue. As discussed earlier, issue resolution requires balanced coordination of policy/procedure, personnel, and technology elements.

## THE NATIONAL PUBLIC SAFETY BROADBAND NETWORK (AKA FIRSTNET)

The National Public Safety Broadband Network (NPSBN) is "a nationwide, standardized, private network with dedicated spectrum to provide public safety access to advanced broadband communications. Once deployed, the NPSBN will enable public safety communications to leverage commercial broadband standards, technologies, devices, and innovations. The NPSBN will also connect to commercial networks and the Internet. Underlying this network will be next generation network (NGN) infrastructure that is converging to packet-switching technology for all forms of communication."[22] Fig. 8.3 provides the overall NPSBN architecture.

There is some overlapping of the communications requirements for the NPSBN and NS/EP (Fig. 8.4). Some capabilities like dynamic prioritization and QoS, however, are unique to the NPSBN. These unique requirements can be fully implemented on the NPSBN but may not be (fully) implemented on commercial networks. Public safety users roaming onto commercial networks will need to be aware of these differences and adjust their communications activities accordingly.

The development of the NPSBN provides the US Government an unprecedented opportunity to coordinate and align policies, requirements, and standards in order to enable innovation, create economies of scale, and ensure that both NS/EP and public safety users' unique communications requirements can be met. A National Security Telecommunications Advisory Committee (NSTAC) report on the NPSBN recommended rationalizing NS/EP and public safety organizations and functions, updating and aligning policies, directing technical initiatives that can support both NPSBN and NS/EP, requiring reporting to facilitate implementation, and addressing funding gaps. Many of these recommendations have not yet been addressed.

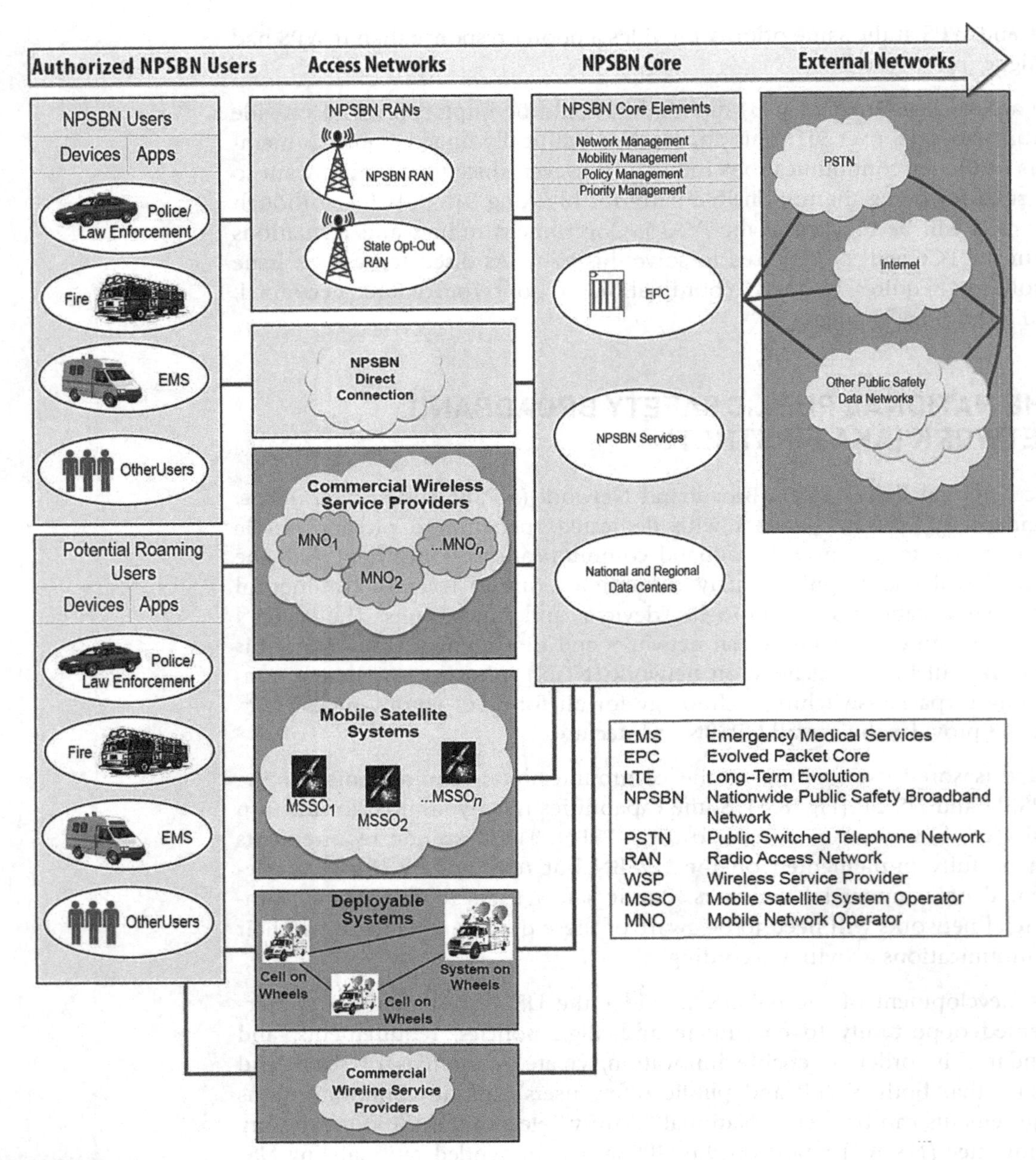


**FIGURE 8.3**
Notional NPSBN diagram depicting ecosystem.[23]

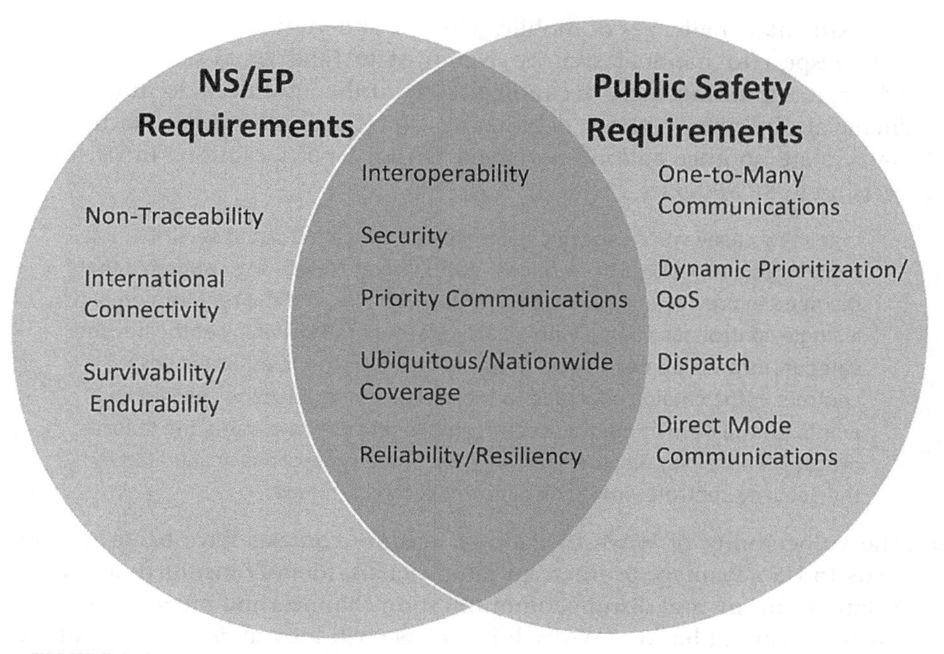


**FIGURE 8.4**
Notional diagram of several NS/EP and public safety functional requirements.[24]

## NATIONAL SECURITY: REAL-WORLD ATTACK SCENARIOS

The challenges of balancing elements like equipment specifications, policies and procedures, and security mechanisms are somewhat different for members of the national security service community than for the emergency management and public safety community members, at least theoretically, because of mandated compliance with FISMA (signed into law as part of the Electronic Government Act of 2002). Policies and procedures, including those that guide product procurement, are more standardized and decision making is more centralized through federal agencies than at the level of state and local governments, in which jurisdictional boundaries are legal and geographical, in addition to being organizational. The differences are converging over time as FISMA and other federal requirements are pushed down to state and local government entities—and manufacturers, consultants, and other business or supply chain partners. Some sourcing practices can be worrisome when applied to national security procurement decisions. For a period of time, as recently as the mid-2000s, 100% of the US Navy's circuit boards were sourced from China. Vulnerabilities can be introduced at lower levels of the OSI stack and cost should not drive procurement decisions.

The particular challenges of mobile, wireless, and VOIP deserve special notice with respect to national security. Nation-state (and quasi-state) actors are adept at exploiting these technologies to further political, territorial, and financial objectives without debilitating concern about whether such activities violate another nation's sovereign territory. NIST cautions in its fourth revision to *SP 800-53*:

> In the few cases where specific technologies are called out in security controls (e.g., mobile, PKI, wireless, VOIP), organizations are cautioned that the need to provide adequate security goes well beyond the requirements in a single control associated with a particular technology. Many of the needed safeguards and countermeasures are obtained from the other security controls in the catalog allocated to the initial control baselines as starting points for the development of security plans and overlays using the tailoring process. There may also be some overlap in the protections articulated by the security controls within the different control families.[25]

The vulnerability of WAPs in national security contexts have been used for years to US advantage to intercept radio signals, identify insurgent and combatant locations, and disrupt communication channels and messages. For this discussion about hacking WAPs, however, we will look at examples of attacks against national governments friendly to the United States.

Russia's attacks in 2012 on mobile phones in the Crimea appear to have been a precursor to physical strikes and perhaps an early warning that Russia was coming. By disrupting Internet service, Russia was able to deny citizen access to the Ukrainian government website. After Crimea's annexation (March 18, 2014), attacks continued against public and private organizations in Poland and the Ukraine (and also included the European Parliament and the European Commission).[26] Reports suggest that Russia had started blocking mobile phone service about three weeks earlier, possibly by jamming wireless signals (phone and radio) using equipment maintained on board its navy ships that were in the Sevastopol port. A few days later (February 28, 2014), several offices of the Ukrainian Ukrtelecom offices were taken over kinetically by individuals who cut communications lines.[27] This example shows that the convergence of kinetic and cyberattack techniques can be effective for those seeking political and territorial gain as well as for those seeking financial and reputational gain.

The tactics in the Ukraine are similar to the one-two, cyber-to-physical punch Russia delivered to Georgia (summer of 2008), in which conventional assault (by land, sea, and air) followed three weeks after cyberattacks were launched against 54 Georgian websites related to communications, finance, and the government.[28] Both Georgia and the Ukraine were perhaps more vulnerable to sustained cyberattack than was Estonia during the attack it experienced in April 2007. At that time, Estonia had earned a reputation as the "most wired country in Europe." More than 96% of its banking transactions were completed online,[29] for example, and its government operations model, according to then-IT director at the Estonian Defense Ministry Mikhel Tammet, was "paperless government."[30] By virtue of its dense Internet coverage and communications network, taking itself offline during the cyber siege, and routing Internet traffic around (not through) Russian territory, Estonia weathered the three week attack well. At the same time, citizens were at a disadvantage with respect to obtaining timely news coverage about what was indeed happening.

As troubling as is Russia's use of cyberwar tactics as a complement to conventional warfare tactics, the takedown of Ukrainian critical power infrastructure in December 2015 may be even more disturbing because of its implication for what the Russian-developed BlackEnergy malware (or other such sophisticated software regardless of its original source) is capable of doing outside Russia's immediate geographical sphere of influence. Initially built as crimeware related to botnets and DDoS attacks (this Trojan Horse is related to a C&C attack against bank cards in the United States between 2010 and 2013[31]), it has since been refined to cripple ICS and SCADA systems via payload delivered in compromised MS Word or Excel documents. BlackEnergy's upgrade includes a backdoor SSH utility, which provides permanent access for attackers.[32] According to Kaspersky Labs, "The BlackEnergy APT group is active in the following sectors: ICS, energy, government, and media in Ukraine; ICS/SCADA companies worldwide; energy companies worldwide."[33]

The capability of taking a sovereign nation's power utility infrastructure down is a game changer. Napoleon observed in the 19th century that "an army marches [or travels] on its stomach. The 21st century version might well be that "an army travels on its internet connectivity."

## TAKEAWAYS

- As part of disaster recovery planning, organizations should consider the impact of limited or no communications capability on their recovery efforts. From a personal perspective, when living in San Francisco, I identified with my daughter and friends where to meet after an earthquake if communications were out. Similar considerations should be taken for mission-critical personnel and functions, and priorities established for critical communications during emergencies.
- Globally, we need definitions about what constitutes "fair play" and what becomes an act of war or a crime against humanity. Essential electrical power and other utilities for hospitals and other medical

facilities, for example, should be protected from wrongful acts by outside parties (even nation-states). "It's time to start defining rules like 'Don't turn off the civilian power grid so hospitals and emergency services can continue'."[34]

- We need effective, active, public-private partnerships to build defensive measures and resilience into our critical infrastructure industries. All levels of government must be engaged, including municipalities, to promote a cooperative, nonadversarial relationship with privately owned utilities.
- Promote more threat intelligence sharing throughout economic sectors using platforms like Structured Threat Information eXpression (STIX, "a structured language for cyber threat intelligence") and Trusted Automated eXchange of Indicator Information (TAXII): international, free, and "community-driven technical specifications designed to enable automated information sharing for cybersecurity situational awareness, real-time network defense and sophisticated threat analysis."[35]
- The best practices mentioned for other use case scenarios apply here as well:
  - Administrative OS and network-based measures
  - Security controls and vulnerability assessment/patch management systems
  - Application control
  - Whitelisting-based controls
  - Cybersecurity awareness training (educating your staff) with particular mention about email-based spear-phishing.

## ENDNOTES


1. Benjamin Mueller (June 2014), "The Laws of War and Cyberspace: On the Need for a Treaty Concerning Cyber Conflict," *London School of Economics Strategic Update 14.2.* Retrieved from http://www.lse.ac.uk/IDEAS/publications/reports/pdf/SU14_2_Cyberwarfare.pdf.
2. National Security Telecommunications Advisory Committee (22 May 2013), "NSTAC Report to the President on the National Security and Emergency Preparedness Implications of a Nationwide Public Safety Broadband Network," p. 10, *NSTAC.* Retrieved from https://www.dhs.gov/sites/default/files/publications/npsbn-final-report-05.
3. US Department of Homeland Security (December 2008), "National Incident Management System (NIMS)," *DHS.* https://www.fema.gov/pdf/emergency/nims/NIMS_core.pdf.
4. https://www.fema.gov/national-incident-management-system.
5. US Department of Homeland Security (February 2016), "Land Mobile Radio (LMR) 101," *DHS/Safecom.* Retrieved from https://www.dhs.gov/sites/default/files/publications/LMR%20 101_508FINAL.pdf.
6. Lynette Luna (1 May 2005), "Unclogging the Grant Pipeline," *IWCE's Urgent Communications.* Retrieved from http://urgentcomm.com/mag/unclogging-grant-pipeline.
7. National Telecommunications and Information Administration (n.d.), "Public safety," *NTIA Website.* Retrieved from https://www.ntia.doc.gov/category/public-safety.

8. Phil Goldstein (20 January 2016), "FirstNet's RFP Could Lead to Spectrum Bonanza for Public Safety Wireless Vendors," *FedTech Magazine*. Retrieved from http://www.fedtechmagazine.com/article/2016/01/firstnet-s-rfp-could-lead-spectrum-bonanza-public-safety-wireless-vendors.
9. Nokia (2014), "LTE Networks for Public Safety Services," *Nokia Networks White Paper.* Retrieved from networks.nokia.com/sites/.../nokia_lte_for_public_safety_white_paper.pdf.
10. US Telecom (February 2016), "The Broadband Internet Economy is Thriving," *US Telecom White Paper.* Retrieved from https://www.ustelecom.org/sites/default/files/files/USTelecom-White-Paper-1.pdf.
11. Ibid., p. 3.
12. Natasha Lomas (19 January 2016), "FCC Urged to Rein in Broadband Providers on Privacy Grounds," *TechCrunch.* Retrieved from https://techcrunch.com/2016/01/19/fcc-urged-to-rein-in-broadband-providers-on-privacy-grounds/.
13. CyberStorm is a cybersecurity exercise initiative composed of several different phases, "a Homeland Security Exercise and Evaluation Program (HSEEP) Tier II exercise focusing on federal strategy and policy." Participants include US Cabinet-level departments and states, international partners, and private sector companies. DHS webpage entitled "CyberStorm: Securing Cyber Space." Retrieved from https://www.dhs.gov/cyber-storm.
14. Findings mentioned in this chapter were contained in the US Department of Homeland Security's "CyberStorm I: Final report" (12 September 2006), "CyberStorm II: Final report" (July 2009), "CyberStorm III: Final report," (July 2011), and "CyberStorm IV: Final report" (June 2015). *DHS.* Retrieved from https://www.dhs.gov/publication/cyber-storm-final-reports.
15. Ronald Timmons (February 2007), "Interoperability: Stop Blaming the Radio," *Homeland Security Affairs* 3, Article 5. Retrieved from https://www.hsaj.org/articles/153.
16. Michael Mandel (March 2016), "Long-Term U.S. Productivity Growth and Mobile Broadband: The Road Ahead," *Progressive Policy Institute Memo.* Retrieved from http://www.progressisvepolicy.org/wp-content/uploads/2016/03/2016.03-Mandel_Long-term-US-Productivity-Growth-and-Mobile-Broadband_The-Road-Ahead.pdf.
17. This is a summarized list. More thorough discussion of the mobile app vetting promise is in Steve Quirolgico, Jeffrey Voas, Tom Karygiannis, Christoph Michael, and Karen Scarfone (January 2015), "Vetting the Security of Mobile Applications," *NIST SP 800-163.* Retrieved from http://nvlpubs.nist.gov/nistpubs/SpecialPublications/NIST.SP.800-163.pdf.
18. The White House (6 July 2012), "Assignment of National Security and Emergency Preparedness Communications Functions," *Executive Order 13618.* Retrieved from https://www.whitehouse.gov/the-press-office/2012/07/06/executive-order-assignment-national-security-and-emergency-preparedness-.
19. This is needed because GSM and CDMA networks are being retired so the spectrum can be more effectively utilized by LTE.
20. Professional best practices do not preclude news reporters from being susceptible to the human proclivity to fill in knowledge gaps with unverified possibility or assumption that includes the belief of omnipotent federal government. This tendency can be used for leverage in marketing, propaganda, psyops, and other disinformation campaigns.
21. For example, Virginia's Fairfax County PSAP has approximately 15 people on staff during peak hours to handle calls from its 1.2 million residents.
22. National Security Telecommunications Advisory Committee, op. cit.
23. Ibid., p. 6.
24. Ibid., p. 14.
25. Joint Task Force Transformation Initiative (April 2013), "Security and Privacy Controls for Federal Information Systems and Organizations," *NIST SP 800-53.* Retrieved from http://nvlpubs.nist.gov/nistpubs/SpecialPublications/NIST.SP.800-53r4.pdf. The control families are access control (AC), audit and accountability (AU), awareness and training (AT), configuration management (CM), contingency planning (CP), identification and authentication (IA), incident response (IR), maintenance, (MA), media protection (MP), personnel security (PS), physical and environmental

protection (PE), planning (PL), program management (PM), risk assessment (RA), security assessment and authorization (CA), system and communications protection (SC), system and information integrity (SI), and system and services acquisition (SA).

26. Owen Matthews (7 May 2015), "Russia's Greatest Weapon may be its Hackers," *Newsweek.* Retrieved from http://www.newsweek.com/2015/05/15/russias-greatest-weapon-may-be-its-hackers-328864.html.
27. Shane Harris (3 March 2014), "Hack Attack," *Foreign Policy.* Retrieved from http://foreignpolicy.com/2014/03/03/hack-attack/.
28. David Hollis (6 January 2011), "Cyberwar Case Study: Georgia 2008," *Small Wars Journal.* Retrieved from http://smallwarsjournal.com/blog/journal/docs-temp/639-hollis.pdf.
29. Jason Richards (n. d.), "Denial-of-Service: The Estonian Cyberwar and its Implications for U.S. National Security," *International Affairs Review.* Retrieved from http://www.iar-gwu.org/node/65.
30. Stephen Herzog (Summer 2011), "Revisiting the Estonian Cyber Attacks: Digital Threats and Multinational Responses," *Journal of Strategic Security,* Vol. 4, No. 2. Retrieved from http://scholarcommons.usf.edu/cgi/viewcontent.cgi?article=1105&context=jss.
31. Owen Matthews, op. cit.
32. Dan Good (4 January 2016), "First Known Hacker-Caused Power Outage Signals Troubling Escalation," *ArsTechnica.* Retrieved http://arstechnica.com/security/2016/01/first-known-hacker-caused-power-outage-signals-troubling-escalation/.
33. Kaspersky Lab (n. d.), "BlackEnergy Attacks in Ukraine," *Kaspersky.* Retrieved from http://www.kaspersky.com/internet-security-center/threats/blackenergy.
34. Daniel Riedel (9 March 2016), "Why We Need to Take Ukraine BlackEnergy Seriously," *Federal Times.* Retrieved from http://www.federaltimes.com/story/government/solutions-ideas/2016/03/08/ukraine-blackenergy-cyberattack/81477758/.
35. US-CERT website on information sharing specifications for cybersecurity. Retrieved from https://www.us-cert.gov/Information-Sharing-Specifications-Cybersecurity.

CHAPTER 9

# Summary and Call to Action

People go through life performing a series of actions (e.g., visiting Grandma) and inactions[1] (e.g., staying home). The benefits and rewards of chosen actions are typically well understood (e.g., enjoying time with Grandma). The risks associated with chosen actions are typically less well understood (e.g., encountering a wolf in the woods on the way to Grandma's house). This is because there are levels of risk (e.g., encountering wolf, getting eaten by wolf) with associated probabilities (e.g., 100% of meeting wolf in forest in fairy tales). If a risk is encountered, a person may perform a contingency action (e.g., drop basket of food for wolf to eat while running away) to escape serious harm. Security consultants call these contingency actions risk mitigation, and the overall process of determining the risks and rewards of certain activities, risk management. People subconsciously perform risk management continually, as it proves beneficial for the survival of the human race.

The goal of this book is to help you, the reader, understand the risks and possible mitigations associated with using mobile devices in your personal and business environments. Hacks are continually evolving, forcing updates to mitigation approaches that combat the hacks. The number of WAPs and mobile devices are increasing (with some devices acting as both access points and mobile devices) as the IoT, self-driving cars, and self-flying drones become a reality. New hacks will be created for these new environments. Because of this, mobile device risk management is an ongoing journey, not a destination.

Taking a pessimist approach and assuming that the good guys can never win so why try and invest in security assures that the bad guys will win. Assuming that with a sufficient budget you will be completely secure (the Big Budget Theory) is overly optimistic; the bad guys will occasionally win as data breaches or network compromises at Target, Sony, Chase Bank, the Federal Office of Personnel Management (OPM), and the Ukrainian Government indicate. Thus one needs to secure personal and business mobile devices as well as possible given time and budget constraints. Since security is a journey and not a destination, one's goal is to be more



Hacking Wireless Access Points. DOI: http://dx.doi.org/10.1016/B978-0-12-805315-7.00009-7

secure tomorrow than today, and more secure today than yesterday. The message, the call to action, is to get started. In Harry S Truman's legendary counsel, "imperfect action is better than perfect inaction."

Protecting against cracking, hacking, skyjacking, juice jacking means starting with the end in mind. Of course, there are at least two ends in competition here: one to protect and preserve assets, and the other to compromise and capture assets. For this reason, it is important to think like a potential attacker and work backwards, rather than getting caught up in the catchy names for different hacking techniques that will continue to proliferate. Successful attack modi operandi will beget others—just as imperfections in tools to detect, disrupt, delay, or deny attacks will beget other tools. Full employment on either side of the security good guy/bad guy divide would seem assured (especially given human frailty and unmerited optimism).

## THINKING LIKE A HACKER: ALIGNING WITH ATTACK METHODOLOGIES

Money is not a reasonable excuse for not having a security program. There are low-cost efforts that can be done to improve security, which is a journey, not a destination. A simple way to get started is to think PEAR: preparation, execution, awareness, repetition. The first four attack steps correlate loosely with preparation and execution. Awareness is the desired state of being well-informed about assets that are connected in some way to WAPs—and that includes those using the connected assets. Because of constantly changing environmental factors like user adaptation of technology and attacker prowess and targets, repetition of secure practices, even after they become habits, is necessary. When I ask my students to "think like hackers" some are initially uncomfortable … and then they get creative (and, at times, just a bit scary). A discussion of PEAR and standard attack steps follows.

### Step 1: Reconnaissance

Preparation encompasses more than just purchasing the cyber equivalent of an off-the-shelf first aid kit. If this Big Bandage Theory were sufficient, information assurance professionals would be worried about redefining their brand (and perhaps pursuing another career). Preparation involves thinking like an attacker and following the classic attacker methodology. So, step one: reconnaissance. Knowing what you have that you value is just a partial step; knowing what others value will get you closer. As was discussed in Chapter 3, Blurred Edges: Fixed and Mobile Wireless Access Points, attacker motivations vary. Identifying the most likely motivations is a good start toward understanding how to prioritize assets for protection.

One of my favorite illustrations of the failure to understand fully what should be considered an asset is in the series of YouTube videos from penetration (i.e., pen) tester Chris Nickerson and his team, "Tiger Team—The Car Dealer Takedown."[2]

Hired to test their clients' (a luxury car dealership) state-of-the-art security system, the team thwarted passcode-protected touchpad access control devices, surveillance cameras, motion detectors, and traditional, heavy-duty locks to break into the facility and *pwn* (power own) not just one of the gorgeous cars for a round-the-block joy ride—which were, of course, covered by insurance to cover damage or theft—but also the PII and financial records for the HNWI clients of the dealership. The videos show clearly how various attack techniques (physical, network penetration, password breaking, social engineering) can be used to satisfy objectives: The attacker has many choices. It also shows the problem of not recognizing truly irreplaceable assets that should be protected most effectively: client trust, personal identity, detailed financial information.

## Step 2: Scanning

Organizations and individuals often object that they do not have the resources to bring in a professional pen testing team, so that preparation step is moved to the indefinite future (e.g., when required by a prospective customer, mandated by legislation or industry standard, or triggered by an incident). Meanwhile, doing an interactive exercise within the work group can at least help identify top priority assets that must be protected. Then one can determine whether assets are finite—like a car or an original, signed contract—or replicable—like digital content that can be extruded and yet still be in place.

This is also the stage in which vulnerability assessments are performed. This is comparable to attack methodology step two: scanning. As either a hacker or an authorized pen tester would do, staff can be challenged to find weaknesses in the way current physical or digital assets are protected, even if they are not members of the IT or security teams. The whole organization can be involved with a low-key incentive—perhaps a gift card for lunch at a favorite eatery. After all, security really is a team sport, and the more engaged the whole organization is the better.

One can think of this exercise as a kind of group SWOT analysis (an "assumptions analysis" could also be revealing) of your organizational assets. Encourage group members to think around the corner and to consider alternative ways of completing critical business or mission processes. The real value of the Y2K processes that organizations endured in the late 1990s, I believe, was the gain in understanding about information flow and how work was done under standard operating conditions and how it could get done if the standard approach was no longer feasible. This should be part of organizational business continuity/disaster recovery planning anyway: defining what must be done, even if on a reduced basis, for a limited period of time. Of course, the Y2K certification benefits actually tended to go to the outsourced consultants doing the actual work. They were the ones to become well educated (on their client's dime) about the operational environment. That knowledge, although documented in often excruciating detail, too often left with the consultants or was captured in pristine binders, duly shelved: snapshot proof that due diligence and due care were honored and respected by the organization.

## Steps 3 and 4: Access and Escalation

With work and information flow defined by actual work teams—a useful exercise for continuous process and quality improvement, not just for security—one can more easily see who and what (if an automated process) should and do have access to specific information and system resources. The area of access and credentials management is the sweet spot for an attacker (step three: access and escalation) when not controlled effectively. Examining this area offers opportunities for asking why, especially when certain permissions and privileges do not align with an individual's (or department's) role, responsibilities, or "need to know." This may be due to personnel reassignment, privilege creep, a one-time exception that became permanent, or other organizational historical artifacts.

Credential misuse was an attack vector in 63% of the confirmed data breaches examined in the "2016 Verizon Data Breach Investigations Report,"[3] meanwhile 60% of those companies responding to a Rapid7 survey acknowledged that they could not distinguish between compromised and legitimate credentials.[4]

Other attractive attack surfaces to be explored at this stage are ghost IT accounts and shadow IT. An example of the former from my experience was found in a research institution that had enabled guest access accounts for the wireless network and some of its data resources. According to policy, these guest accounts were to be enabled for the duration of a researcher's participation on a grant team, then disabled after the grant work was fulfilled. In practice, however, these accounts were not disabled "just in case" the researcher should need access to them at some time in the distant future (even from the grave, apparently, since at least one of the researchers had already departed this physical world).

Shadow IT is, basically, rogue IT infrastructure. Although it is closely associated with bring your own device practices—as when staff use personal, unsegmented laptops or cell phones for business communications—it can also include unauthorized wireless routers and signal boosters ("I just needed better connectivity") and nonstandard software. Frequently shadow IT is less a symptom of malice on the part of staff and more a symptom of a dysfunctional relationship between the technology and user groups. IT and security teams must be perceived as being an enabler of work group performance and not an impediment.[5]

## Step 4: Exfiltration

Guarding against data exfiltration requires comprehensive knowledge of what data exists; how it should be protected and monitored in its various states (in process, in transit, at rest); who interacts with the data (as creator, user, custodian, owner); where it is located (including media storage format); what is its volatility (frequency changes, declassification schedule, destruction schedule); where is it most vulnerable; and so forth. By adopting the perspective of the typical opportunistic (non-nation-state) attackers, you can identify the data that will be most attractive and the conditions under which that data will be most vulnerable.

## Steps 5, 6, 7: Sustainment, Assault, Obfuscation

### *Incremental Security Actions*

Frequently organizations and individuals object that they are not big enough or financially stable (flush) enough or important enough to deploy security solutions beyond tepid passwords and somewhat up-to-date AV software on the network. Such organizations and individuals promote, in effect, that sorry risk management strategy that no one likes to admit because it sounds (and is) inconsistent with reasonable care: denial. I do believe that denial is practiced regularly as a risk management strategy alongside the prescribed ones: mitigate, transfer, accept, avoid.

To avoid an information and network access free-for-all, WAPs should not be construed—nor configured—as being free for all. Takeaways from Chapter 2, Wireless Adoption, Chapter 3, Blurred Edges: Fixed and Mobile Wireless Access Points, Chapter 4, Hacks Against Individuals, Chapter 5, WAPs in Commercial and Industrial Contexts, Chapter 6, WAPs in Medical Environments, Chapter 7, Hacking Wireless Access Points: Governmental Context, and Chapter 8, Non-Civilian Government Context, are summarized below under general categories for individuals, organizations, and automated controls systems (potentially relevant to individuals and organizations). Their order of presentation does not indicate recommended priorities: Individuals and organizations should make their own decisions about where to start. Multiple objectives are addressed within these takeaways; guaranteed, hack-free security is not one of them.

The first objective for implementing these practices is to raise the level of cost and inconvenience to the attacker so that his or her attack cost/benefit calculation is no longer attractive. The second objective is to mitigate the effects of loss or compromise of assets so that one is not vulnerable to ransomware demands, for example, or so that protected information is rendered practically unusable (i.e., is encrypted). The third objective is to encourage shared responsibility, whether across all organizational groups (not just IT or security teams), with device manufacturers (e.g., build security in, force changes to default passwords), or throughout the business partner network and supply chain. The fourth objective is to share information about potential issues (e.g., possible precursor activity, IoC, IoA, product or service vulnerability) and report incidents to appropriate clearinghouses (e.g., National Center for White Collar Crime <https://www.nw3c.org/>). The fifth objective is to choose good WAP hygiene over convenience and encourage others to do so.

## CALL TO ACTION FOR ORGANIZATIONS

Essential considerations for reducing one's WAP vulnerability profile are least privilege access to network resources, network segmentation, user and device authentication, application control (including patch and default setting maintenance), contractually defined secure supply chain, data protection, incident response and recovery. The following is a kind of checklist of recommended actions for different components of a wireless communications system.

### Mobile and Fixed Devices

- Secure loaner and dedicated mobile devices for remote travel use with secure VPN connectivity (1024-bit encryption or better)
    - Examine devices upon return for malware presence
    - Revise security functions as deemed necessary
    - Wipe and re-image the device for future use
- Provision secure mobile devices for a single, VPN-only connection to the agency portal, which then manages connectivity to Internet resources (per US Government secure mobile computing initiative)

- Control fixed WAP communications including peripheral devices (e.g., printers)
  - Restrict access to whitelisted uses
  - Implement white and black lists of executable files
- Change manufacturer or service provider default settings and passwords
  - Implement passwords/codes that are not obvious or easily guessable
  - Create unique administrator accounts
  - Disable unnecessary services permanently and enable intermittently necessary services (e.g., Bluetooth) only when needed
- Encrypt protected data on endpoint devices

### Network Architecture

- Layer digital and analog control mechanisms to reduce attractiveness as a target
  - Implement air gaps between digital devices in commercial facilities
  - Require physical access to mechanical systems for administration
  - Harden remote access (even by legitimate administrators) to prevent lateral movement across networks (otherwise, if the administrator's credentials are compromised, your network is)
- Segregate critical communications (consider the following as examples)
  - Medical categories: patient, clinician, procedural (e.g., M2M for remote surgery)
  - Power/utility categories: ICS device control, ICS and SCADA data capture, mechanical system control
  - Military categories: administrator, field combatant, commander
  - Emergency/public safety: dispatch, field response team member, device (e.g., wearable or fixed surveillance cameras) data capture

### Credentials and Access Control

- Isolate, harden, and prudently store credentials (avoid obvious file naming conventions like "passwords" or "access codes")
- Use two factor-authentication or robust passwords (at least 16 digits, alphanumeric, interspersed special symbols, mixed upper and lower case)
- Track activity, including privileged activity, by individual user not user role accounts (e.g., system administrator)
- Monitor scripted processes (business-to-business or M2M) for anomalies

- Limit and segregate external access by vendors (deny lateral movement across network)
  - Require that vendors with system access have compatible security policies and practices
  - Impose contractual requirements regarding vendor responsibility for system compromise (increase cost of inaction)

### Data Protection

- Encrypt protected data assets that are accessible over networks
- Store archival documents in repositories not accessible through web server gateways

### Security-Conscious Culture[6]

- Require interactive awareness training at least annually and after major incidents (e.g., merger/acquisition, ERP implementation, system compromise, new hire, major reduction in force)
- Train public-facing staff to recognize, handle, and report social engineering attempts[7]
- Map information and process flows

### Incident Response

- Develop, test, practice, and document plans for cyber or physical incidents
- Define call trees and communications plans with key stakeholders

## CALL TO ACTION FOR INDIVIDUALS

Individual users should consider implementing restricted physical access to devices, trusted communications channels, hardened home-based infrastructure, and off-device data storage. Basic recommended checklist items for an individual's wireless communications system are listed below.

### Mobile and Fixed Devices

- Personalize your device(s), including things like wireless routers, printers, surveillance cameras
  - Use a passcode or password that is difficult to guess
  - Segregate business from personal use
  - Manage your apps for information-sharing policies and licensed use
  - Disable unnecessary services permanently and enable intermittently necessary services (e.g., Bluetooth) only when needed

- Upgrade minimally secure home wireless routers and firewalls to industrial strength
  - Change manufacturer or service provider default passwords
- Question the need for Internet-connected home appliances

### Networks

- Avoid public Wi-Fi hotspots for business and personal financial transactions or confidential information exchanges even when using a VPN
  - Know that even two-factor authentication can be compromised if both pieces of information travel through the same WAP, which neutralizes the protective effect of "out of band" communication
  - Verify the network name with the Wi-Fi owner
  - Remember that rogue WAP with malware embedded into "login" page is easily installed with a deceptive name

### Data Protection

- Back up mobile device information to the cloud and/or off-line storage, then delete protected information that could be exposed if device were lost, stolen, or pwned

## CALL TO ACTION FOR AUTOMATED SYSTEMS

Automated controls system should be characterized by network separation of functions, strong controls of information flows across networks, secure remote communications. Recommended checklist items for wireless communications systems are captured below.

### Devices

- Protect individual ICS components from exploitation
- Restrict logical and physical access to ICS network and devices

### Network Architecture

- Implement redundant WAPs with unique encryption passcodes
  - Protect against latency issues by implementing dedicated network for mission-critical controls
  - Segregate ICS from internal administrative networks
  - Separate both ICS and administrative networks from public access networks over shared WAPs

- Restrict the flow of data between ICS, administrative, and public access networks
- Encrypt remote communications into ICS and administrative networks (will not have negative impact on remote monitoring)

### Data Protection

- Track modification of data and restrict its unauthorized modification
- Implement IDS (IPS can be gamed and inadvertently create a self-imposed DoS)

## THE IMPORTANCE OF THINKING EARNESTLY

We all have choices about how we implement and use technology, making our own personal cost/benefit calculation about our appetite for risk versus convenience and prioritizing actions (or inactions) accordingly. As dorm room posters in the late 1960s posters proclaimed, "Not to decide is to decide."[8] Although we've not touched much on ethical considerations in these chapters, new technologies and new uses for technologies invite such considerations. The ongoing debate about autonomous cars, for example, poses ethical questions about how these vehicles would be programmed to act in situations when it is inevitable that someone will be hurt.[9]

Continuing with the autonomous car example, which is so dependent on wireless technologies as a system of components, we can analyze it from an engineering perspective as described in "How Might a Security Engineer Look at Autonomous Vehicles?"

Making risk calculations requires inviting multiple perspectives—ethical, engineering, business, legal, technological—and thinking earnestly. Security decisions, actions, and inactions require the whole team.

## CONCLUSION

WAPs are everywhere: in or around our pockets, cars, coffee shops, cameras, streetlights. They rely on chatty technology that is often undiscriminating about the intentions of those receptors to which they are transmitting information or allowing access. We can choose to give them up (unlikely answer) or choose to use them responsibly (good answer). We must respect their power and learn to control it to our advantage and not abdicate that control to others. Although security is an incremental, often experimental, process—it's worth the investment. The consequences of failing to act, or failing to think while acting, are too disturbing—whether considered at the level of the individual, organization, or nation-state.

Thank you for taking the time to consider.

## HOW MIGHT A SECURITY ENGINEER LOOK AT AUTONOMOUS VEHICLES?

First, the engineer would look at the system components:

- Highways and streets with smart sensors (e.g., speed sensors, cameras) and C&C center (AKA the smart grid)
- Vehicles without any ability to communicate with the smart grid or other vehicles (e.g., 1970 Mustang)
- Vehicles with an ability to communicate and act in concert with the smart grid and other vehicles (best-case scenario is all vehicles are this way—dream on)
- Vehicles with an ability to communicate with the smart grid and other vehicles but will not act in concert with them. These vehicles include those used in motorcades, fire, police, and emergency management activities, which will be piloted by human beings to get around problems not programmed into the smart grid
- Vehicles with an ability to communicate with the smart grid and other vehicles but will not act in concert with them. In addition, these vehicles have the ability to direct the smart grid and other vehicles. These select vehicles will belong to presidential motorcades, and police and fire commanders

Next the engineer would look at how the components interact:

- How does the smart grid communicate, identify, and track each vehicle on its grid?
- How does each vehicle communicate with the smart grid?
- How does each vehicle communicate, identify, and track each vehicle surrounding it?
- What does the smart grid do when the communications it receives from a vehicle does not match its sensor input?
- What does each vehicle do when the communications it receives from the smart grid does not match its sensor input?
- What does each vehicle do when the communications it receives from other vehicles does not match its sensor input?
- What does each vehicle do when no communications are received from the smart grid?
- How does the smart grid validate and act on commands from "priority" vehicles (e.g., presidential motorcades)?
- How does each vehicle validate and act on commands from "priority" vehicles (e.g., presidential motorcades)?

Interactions must be analyzed for normal conditions and for systems under stress. An example of a stress condition is a city being told it has *X* hours to evacuate before a disaster strikes. In this case, one needs to determine:

- How many people will change their autonomous vehicles into manual mode to "get out of Dodge" as quickly as possible?
- How many people with autonomous vehicles will have well-practiced manual driving skills for maneuvering during a high-stress situation?
- How many people without autonomous vehicles or with vehicles only with autonomous capabilities will go to the smart grid to either get a ride or to highjack a vehicle (see 1953 movie *War of the Worlds*)?
- How efficient is the smart grid in this case?

Next, the engineer needs to think like a criminal to determine:

- How can each of the above components and communications be used in the commission of a crime?
- How can each of the above components and communications be used to escape detection and/or pursuit?

And the engineer needs to think like a terrorist to determine:

- How can each of the above components and communications be used to create fear, uncertainty, and doubt in or terrorize the public?
- How can each of the above components and communications be used to escape detection and/or pursuit?

The engineer looks for countermeasures to protect against criminal and terrorist activity. The countermeasures need to be vetted to ensure they do not adversely impact the functions of the smart grid and autonomous vehicles. Then they need to be vetted to see if and how criminals and terrorists could make use of the countermeasures if they are hacked.

Although some may believe that the above will keep security engineers awake at night, in reality they sleep like a baby knowing that they have employment for as long as they want.

## ENDNOTES

1. In this discussion, inaction is defined as an action of doing nothing.
2. YouTube, "Tiger Team—The Car Dealer Takedown." Retrieved from https://www.youtube.com/watch?v=MdQas_We_kI.
3. Verizon (2016), "2016 Data Breach Investigations Report," *Verizon (and team)*. Retrieved from http://www.verizonenterprise.com/verizon-insights-lab/dbir/2016/.
4. Ms. Smith (13 January 2016), "60% of Companies Cannot Detect Compromised Credentials, Survey says," *NetworkWorld*. Retrieved from http://www.networkworld.com/article/3022066/security/60-of-companies-cannot-detect-compromised-credentials-say-security-pros-surveyed.html.
5. Andrew Froehlich (18 March 2015), "Shadow IT: 8 Ways to Cope," *InformationWeek*. Retrieved from http://www.informationweek.com/strategic-cio/it-strategy/shadow-it-8-ways-to-cope/d/d-id/1319535.
6. There are many resources available on the Internet for training materials. The US Small Business Administration (SBA), for example, has posted a good 30-minute presentation on its website <https://www.sba.gov/tools/sba-learning-center/training/cybersecurity-small-businesses>. The information is practical, solid, and comprehensible.
7. All employees need to understand signs of a social engineering attempt. "The [2016 DBIR] report also revealed the power of a properly socially engineered phishing attack. The data, which was derived from sanctioned phishing tests that had 8 million total results, showed that 30 percent of phishing messages were opened by the target with 12 percent moving on to click the malicious attachment or link. This is up from 2014 when only 23 percent opened the email with 11 percent clicking on the attachment." Doug Olenick (27 April 2016), "Phishing, POS and Stolen Credentials Top Data Breach Methods: Verizon," *SC Magazine*. Retrieved from http://www.scmagazine.com/phishing-pos-and-stolen-credentials-top-data-breach-methods-verizon/article/492641/.
8. Attributed to Harvey Cox, but perhaps borrowing from earlier philosopher Jean-Paul Sartre (mid-20th century) who borrowed from René Descartes (17th century): "If you choose not to decide you still have made a choice."
9. The MIT Technology Review posted the following commentary on June 24, 2016. Retrieved from https://blu185.mail.live.com/ ?tid=cmLypq0RE65hGGOdidZ1yJrg2&fid=flsearch&srch=1&skws=autonomous%20car&sdr=4&satt=0.

   **Not in *My* Autonomous Car, Thank You Very Much** If your self-driving car had to kill you, the driver, or a pedestrian who stepped out in front of it, which should it choose? Philosophers are already grappling with those kinds of questions, but a new series of surveys published in Science today reveals a potentially more important set of opinions on the topic—namely those of the general public that will be buying the cars. There's a strong not-in-my-back-yard vibe to the results: people approve of autonomous vehicles that might sacrifice passengers to save the lives of others—but they wouldn't want to ride in such a vehicle themselves. Of course, turning the situation into a binary decision—us or them—oversimplifies the situation that autonomous cars may find themselves in, where there will almost always be scope to at least try and save the lives of all of the parties involved. But we still need to know how the cars should behave in the worst case. As the Harvard psychologist Joshua D. Greene writes: "Before we can put our values into machines, we have to figure out how to make our values clear and consistent." So far, we seem to be struggling.

# Appendix

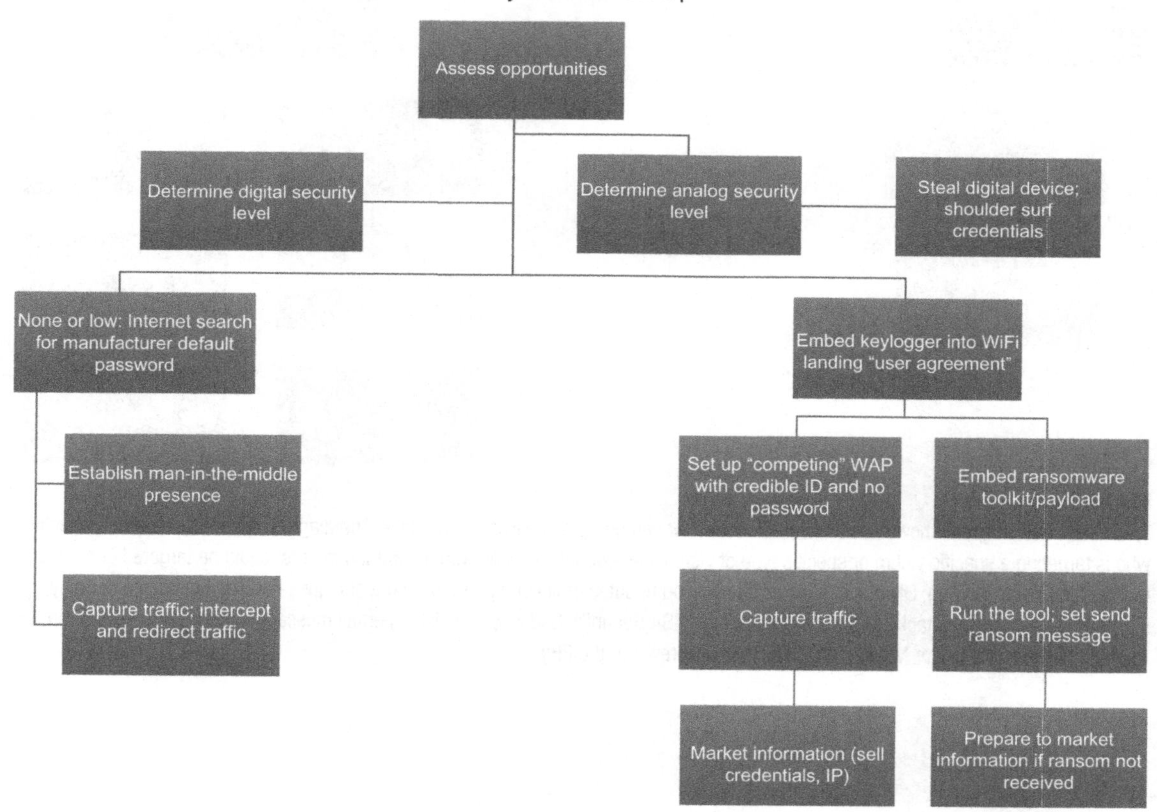

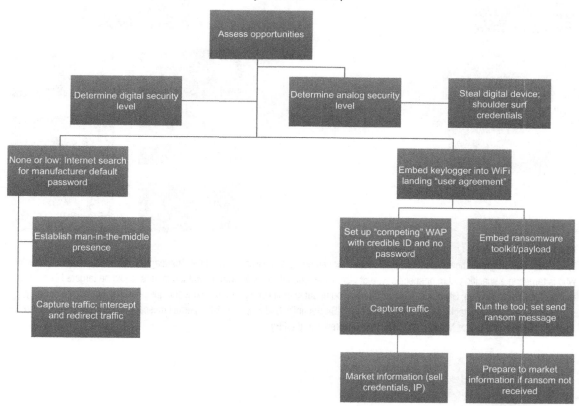


**FIGURE 1.1**
This attack tree diagram shows basic attacker options for compromising a public WiFi hotspot. The diagram assumes an opportunistic attacker who is not targeting a specific victim, although these methods could be combined with others if a specific victim or victim type is being targeted.

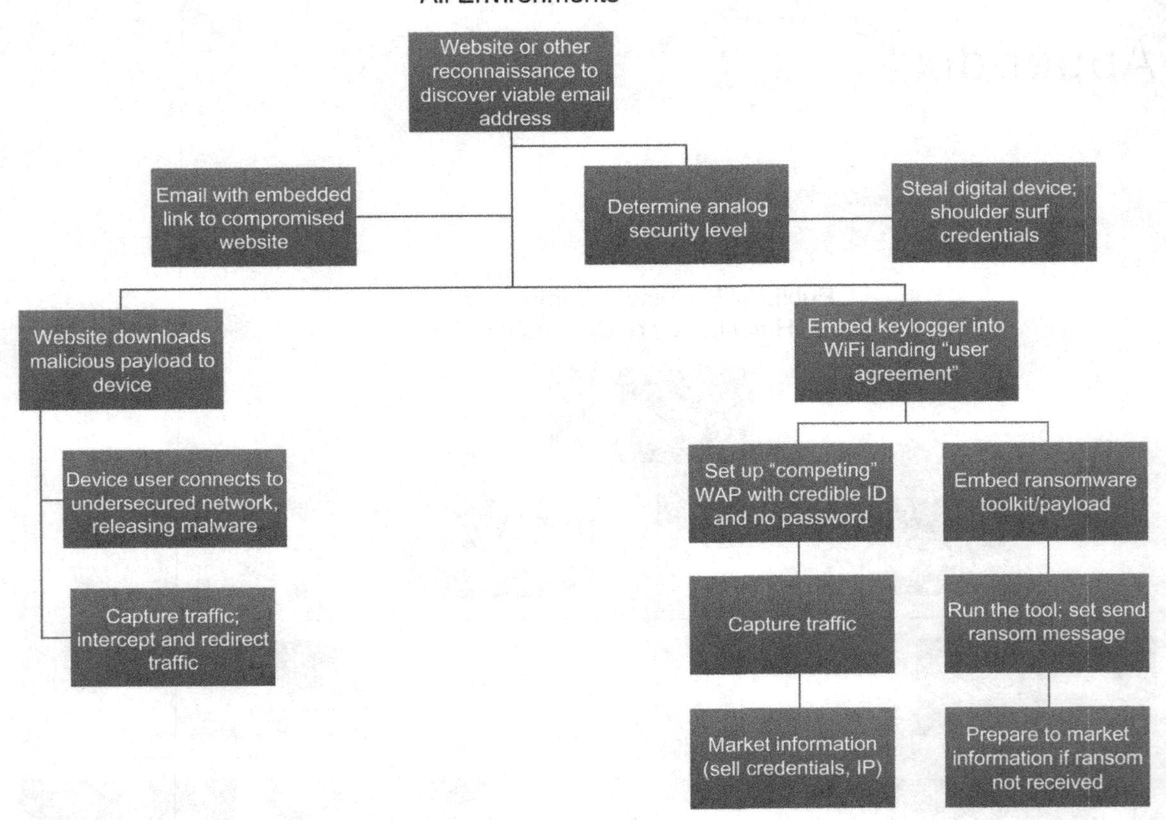


**FIGURE 1.2**

This attack tree diagram shows basic attacker options for setting up a spearphishing attack. The diagram assumes a focused attacker who is targeting a specific victim or specific type of victim. For example, victims with a common interest could be targeted (e.g., sports fans looking for World Cup, Olympics, or other information updates or links). Spearphishing techniques have be been used in many high-profile, successful attacks (e.g., Sony Playstation, RSA SecurID). Sadly, spearphishing email messages even may appear to be from the National Center for Missing and Exploited Children (per the FBI).

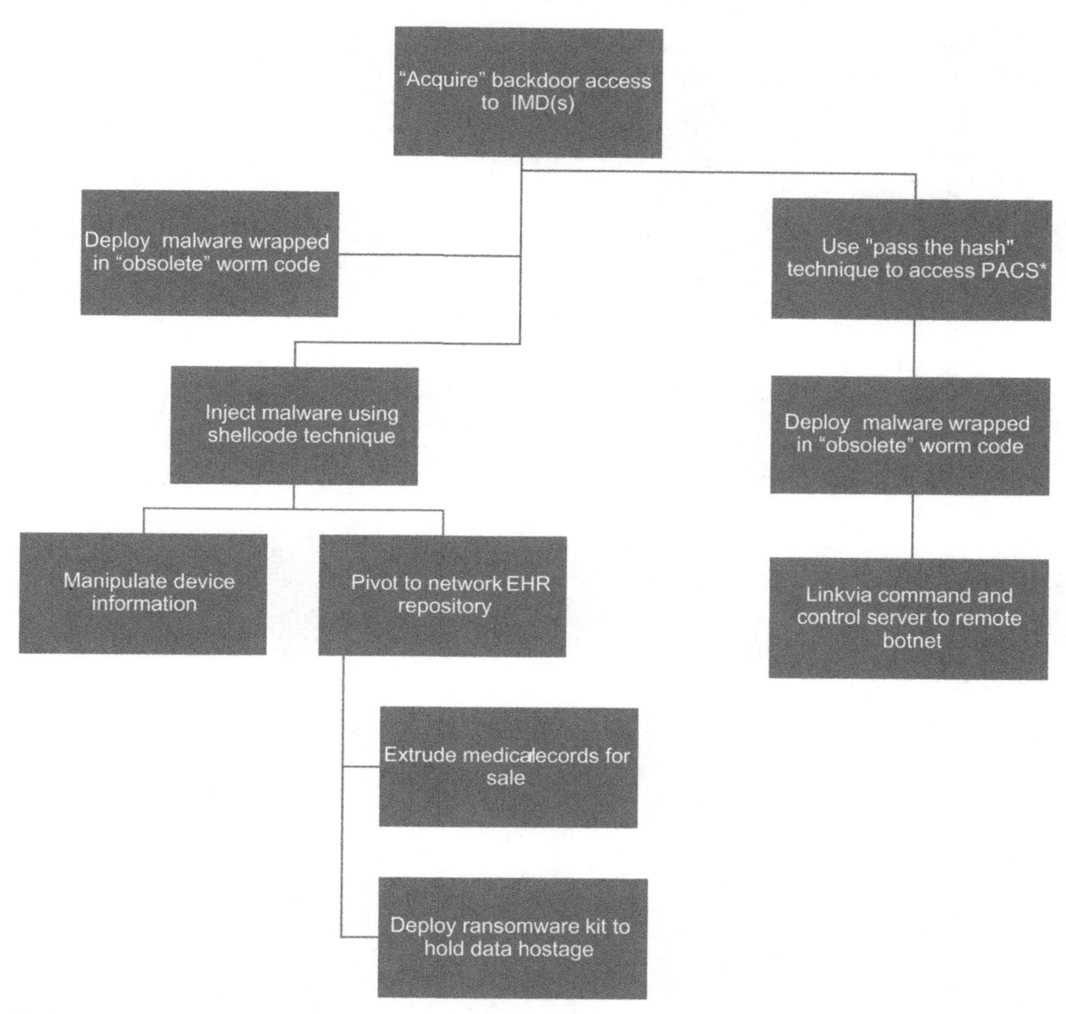


**FIGURE 1.3**

This attack tree diagram shows sample basic attacker options for gaining access to a secured medical environment by first compromising common patient monitoring devices and using them as a back door into the network. Once inside the trusted network (in which instrument traffic is not effectively segregated), the attacker can perform other reconnaissance, capture credentials, deploy malware like ransomware, extrude high-value medical record information (worth 10–20 times more than credit card information), connect to an external command and control server. Many devices, such as blood gas analyzers or infusion pumps, were not designed and manufactured with security in mind and have limited, if any, patching potential.
**PACS*, Picture Archive and Communications System (used for images captured by various devices in different departments.

# Glossary

| Term | Definition | Definition Source |
|---|---|---|
| Active tag | An RFID tag that relies on a battery for power. | NIST SP 800-98. *Guidelines for Securing Radio Frequency Identification (RFID) Systems.* Retrieved from http://csrc.nist.gov/publications/nistpubs/800-98/SP800-98_RFID-2007.pdf. |
| Administrative account | A user account with full privileges on a computer. Such an account is intended to be used only when performing personal computer (PC) management tasks, such as installing updates and application software, managing user accounts, and modifying operating system (OS) and application settings. | Draft NIST Special Publication 800-114, Revision 1. *User's Guide to Telework and Bring Your Own Device (BYOD) Security.* |
| Aircracking | Password-cracking technique used against Wired Equivalent Privacy and Wi-Fi Protected Access (WPA) protections; captures wireless packets to recover password using Fluhrer, Mantin, and Shamir (FMS) attack. | Infosec Institute (23 February 2015), "13 Popular Wireless Hacking Tools." Retrieved from http://resources.infosecinstitute.com/13-popular-wireless-hacking-tools/. |
| Airjacking | Technique for injecting forged packets to support a MITM or DoS attack. | Infosec Institute (23 February 2015). "13 Popular Wireless Hacking Tools." |
| Bluejacking | Technique sends unsolicited, often anonymous, messages over Bluetooth to Bluetooth-enabled devices; messages may contain a vCard (typically for connection to another Bluetooth-enabled device via OBEX protocol; uses include bluedating and bluechatting). | Multiple discussions, for example, those retrieved from Wikipedia https://en.wikipedia.org/wiki/Bluejacking and Professor Messer's Comptia Security+ textbook http://www.professormesser.com/security-plus/sy0-401/bluejacking-and-bluesnarfing-2/. |

| Term | Definition | Definition Source |
|---|---|---|
| Bluesnarfing | Technique to obtain unauthorized access of information from one Bluetooth-enabled device by another. | Multiple discussions, for example, those retrieved from Wikipedia https://en.wikipedia.org/wiki/Bluejacking and Professor Messer's *Comptia Security+* textbook http://www.professormesser.com/security-plus/sy0-401/bluejacking-and-bluesnarfing-2/. |
| Caller ID spoofing | Falsifying the caller ID to a number other than the actual calling station's. | Andrew Swoboda (20 April 2015), "How to Protect Yourself From Caller ID Spoofing," *Tripwire.* Retrieved from http://www.tripwire.com/state-of-security/security-awareness/how-to-protect-yourself-from-caller-id-spoofing/ and Robert Lemos (4 August 2015), "Trust No One: How Caller ID Spoofing Has Ruined the Simple Phone Call," *PC World.* Retrieved from http://www.pcworld.com/article/2951748/security/trust-no-one-how-caller-id-spoofing-has-ruined-the-simple-phone-call.html. |
| Commjacking | Intercepting transmissions between any device and the Wi-Fi or cellular networks to which it is connected. | Ben Kepes (12 May 2015), "CoroNet Launches to Put a Stop to 'Commjacking,'" *Forbes.* Retrieved from http://www.forbes.com/sites/benkepes/2015/05/12/coronet-launches-to-put-a-stop-to-commjacking/. |
| Content filtering | The process of monitoring communications such as email and web pages, analyzing them for suspicious content, and preventing the delivery of suspicious content to users. | Draft NIST Special Publication 800-114, Revision 1. |
| CVV | Credit card industry acronym for card verification value. CVV1 is a unique three-digit value encoded on the magnetic stripe of the credit card. | Charles McFarland, François Paget, and Raj Samani (n. d.), "The Hidden Data Economy: The Marketplace for Stolen Digital Information," *McAfee Labs.* Retrieved from http://www.mcafee.com/us/resources/reports/rp-hidden-data-economy.pdf. |
| CVV2 | The three-digit value printed on the back of the credit card. | McFarland, Paget, and Samani. |

| Term | Definition | Definition Source |
|---|---|---|
| Cyberattack methodology (classic model) | Step 1: Reconnaissance<br>Step 2: Scanning<br>Step 3: Access and escalation<br>Step 4: Exfiltration<br>Step 5: Sustainment<br>Step 6: Assault<br>Step 7: Obfuscation | Identity Week (6 June 2016), "The Seven Steps of a Successful Cyber Attack," *Identity Week*. Retrieved from https://www.identityweek.com/seven-steps-of-successful-cyber-attack/. |
| Diameter (as in diameter signaling) | One of two signaling protocols that drive most of communications in IMS networks (the other is session initiation protocol, SIP). Diameter is the industry standard for data signaling from mobile devices such as smartphones and tablets. The two protocols are designed to perform separate but complementary functions in IMS/LTE networks. For example, when a SIP session is initiated, diameter messages are working behind the scenes within the core network to authenticate that the subscriber is who they say they are, is authorized to use certain network services or applications, and is charged correctly for using those services. SIP is the industry standard for message signaling in real-time communications such as Voice over IP (VoIP) and videoconferencing sessions. | http://www.sonus.net/sites/default/files/The%20ABCs%20of%20Diameter%20Signaling_0.pdf. |
| Doxing (or doxing) | Derived from an abbreviation for document; the process of making another person's identity by researching social media artifacts, job applications, etc. The underlying motivation for response can range from benign (e.g., as part of a pre-interview search or AirBnB reassurance) to malicious (e.g., reconnaissance to discover privileged access credentials or financial information). | Cyberbullying Research Center (16 September 2015), "Doxing and Cyberbullying," *Cyberbullying Research Center.* Retrieved from http://cyberbullying.org/doxing-and-cyberbullying. |
| Drone skyjacking | Drone engineered to take control of other drones within wireless or flying range. | Samy Kamkar (2 December 2013), "SkyJack," *Applied Hacking* [Website]. Retrieved from https://samy.pl/skyjack/. |

| Term | Definition | Definition Source |
|---|---|---|
| Fullzinfo | Information "package" in which the seller supplies all of the details about the credit card and its owner, such as full name, billing address, payment card number, expiration date, PIN number, social security number, mother's maiden name, date of birth, and CVV2. | Charles McFarland, François Paget, and Raj Samani (n. d.), "The Hidden Data Economy: The Marketplace for Stolen Digital Information," *McAfee Labs.* Retrieved from http://www.mcafee.com/us/resources/reports/rp-hidden-data-economy.pdf. |
| Fullzinfo with COB | Refers to those credit cards with associated login and password information. Using these credentials, the buyer can change the shipping or billing address or add a new address. | MacFarland, Paget, Samani. |
| Google dorking | Use of advanced operators or complex search strings to discover sensitive, specific information about an organization or individual. Associated, for example, with early 2016 hack into a New York water control system. | Srinivas Mukkamala (9 June 2016), "Google Dorking: Exposing the Hidden Threat," *InformationWeek.* Retrieved from http://www.darkreading.com/cloud/google-dorking-exposing-the-hidden-threat/a/d-id/1325842. |
| ISM | Acronym for industrial, scientific, and medical, a reference to radio bands or parts of the radio spectrum that are usually available for use for any purpose without a license. | PC Magazine (n.d.), "ISM Band." Retrieved from Retrieved from http://www.pcmag.com/encyclopedia/term/45467/ism-band. |
| Passive tag | A tag that does not have its own power supply. Instead, it uses RF energy from the reader for power. Due to the lower power, passive tags have shorter ranges than other tags, but are generally smaller, lighter, and cheaper than other tags. | NIST SP 800-98. |
| Personal firewall | A software program that monitors communications between a computer and other computers and blocks communications that are unwanted. | Draft NIST Special Publication 800-114, Revision 1. |
| Popup window | A standalone web browser pane that opens automatically when a web page is loaded or a user performs an action designed to trigger a popup window. | Draft NIST Special Publication 800-114, Revision 1. |

| Term | Definition | Definition Source |
|---|---|---|
| Semiactive tag | A tag that uses a battery to communicate but remains dormant until a reader sends an energizing signal. Semi-active tags have a longer range than passive tags and a longer battery life than active tags. | NIST SP 800-98. |
| Semipassive tag | A passive tag that uses a battery to power on-board circuitry or sensors but not to produce back channel signals. | NIST SP 800-98. |
| Skyjacking | Exploiting over-the-air provisioning (OTAP) protocols to trap wireless access points into connecting to a rogue wireless LAN controller (WLC) or access point; works by transmitting fake radio resource management (RRM) messages with information about the fake WLC; supporting tools include packet injection software. | AirTight Networks (n.d.) [Webinar]. Retrieved from http://www.airtightnetworks.com/fileadmin/pdf/webinar/Skyjacking_FAQs.pdf. |
| Smart dust | "Miniaturized sensors/transmitters that are sprinkled onto an area such as a battlefield and used to analyze the environment. Developed by Professor Kris Pister at the University of California at Berkeley and expected in the next decade, smart dust particles are planned to be no more than one cubic millimeter in size, which includes a solar cell, a sensor, CPU, memory and radio transmitter." | *PC Magazine Encyclopedia*. Retrieved from http://www.pcmag.com/encyclopedia/term/51508/smart-dust. |
| Smart dust mote | A commercial wireless networking device for ultra-low-power, self-configuring, self-healing wireless mesh networks. | J. Roger Bowman and Darrin Wahl (June 2012), "Advanced Distributed Sensor Networks for Electric Utilities," *SAIC Report for California Energy Commission (CEC-500-2012-069)*. Retrieved from http://www.energy.ca.gov/2012publications/CEC-500-2012-069/CEC-500-2012-069.pdf. |

| Term | Definition | Definition Source |
|---|---|---|
| Software-generated credit card | A valid combination of a primary account number (PAN), an expiration date, and a CVV2 number that has been generated by software. Valid credit card number generators can be purchased or found for free online. As these tools can be easily found, their generated combinations do not have market value. | MacFarland, Paget, Samani. |
| Whitelist | In messaging terms, a list of email senders that are known to be benign, such as coworkers, friends, and family. A user can add their email addresses to a whitelist, which will cause their future emails to not be classified as spam. Spam filtering accidentally classifies some emails as spam that are not, but a whitelist overrides that classification and ensures that emails from trusted senders are received by the user. Whitelist also refers to a list of websites, applications, etc. deemed acceptable. (Not surprisingly, the opposite of a *black list.*) | Draft NIST Special Publication 800-114, Revision 1. |
| WISP | Acronym for Wireless Internet Service Provider. | IEEE documentation. |

# Index

*Note*: Page numbers followed by "*f*" and "*t*" refer to figures and tables, respectively.

## H



## I



## J



## K



## L



## M

Printed in the United States
By Bookmasters